Popular Shakespeare

Palgrave Shakespeare Studies

General Editors: Michael Dobson and Gail Kern Paster

Editorial Advisory Board: Michael Neill, University of Auckland; David Schalkwyk, University of Capetown; Lois D. Potter, University of Delaware; Margreta de Grazia, Queen Mary University of London; Peter Holland, University of Notre Dame

Palgrave Shakespeare Studies takes Shakespeare as its focus but strives to understand the significance of his oeuvre in relation to his contemporaries, subsequent writers and historical and political contexts. By extending the scope of Shakespeare and English Renaissance Studies the series will open up the field to examinations of previously neglected aspects or sources in the period's art and thought. Titles in the *Palgrave Shakespeare Studies* series seek to understand anew both where the literary achievements of the English Renaissance came from and where they have brought us.

Titles include:

Pascale Aebischer, Edward J. Esche and Nigel Wheale (*editors*)
REMAKING SHAKESPEARE
Performance across Media, Genres and Cultures

Mark Thornton Burnett
FILMING SHAKESPEARE IN THE GLOBAL MARKETPLACE

David Hillman
SHAKESPEARE'S ENTRAILS
Belief, Scepticism and the Interior of the Body

Jane Kingsley-Smith
SHAKESPEARE'S DRAMA OF EXILE

Stephen Purcell
POPULAR SHAKESPEARE
Simulation and Subversion on the Modern Stage

Paul Yachnin and Jessica Slights
SHAKESPEARE AND CHARACTER
Theory, History, Performance, and Theatrical Persons

Forthcoming titles:

Timothy Billings
GLOSSING SHAKESPEARE

Erica Sheen
SHAKESPEARE AND THE INSTITUTION OF THE THEATRE

Palgrave Shakespeare Studies
Series Standing Order ISBN 978-1403-911643 (hardback) 978-1403-911650 (paperback)
(*outside North America only*)

You can receive future titles in this series as they are published by placing a standing order. Please contact your bookseller or, in case of difficulty, write to us at the address below with your name and address, the title of the series and the ISBN quoted above.

Customer Services Department, Macmillan Distribution Ltd, Houndmills, Basingstoke, Hampshire RG21 6XS, England

---

Popular Shakespeare

Simulation and Subversion on the Modern Stage

Stephen Purcell

palgrave
macmillan

First published 2009 by
PALGRAVE MACMILLAN

Palgrave Macmillan in the UK is an imprint of Macmillan Publishers Limited,
registered in England, company number 785998, of Houndmills, Basingstoke,
Hampshire RG21 6XS.

Palgrave Macmillan in the US is a division of St Martin's Press LLC,
175 Fifth Avenue, New York, NY 10010.

Palgrave Macmillan is the global academic imprint of the above companies
and has companies and representatives throughout the world.

Palgrave® and Macmillan® are registered trademarks in the United States,
the United Kingdom, Europe and other countries.

ISBN-13: 978-0-230-57703-9   hardback

This book is printed on paper suitable for recycling and made from fully
managed and sustained forest sources. Logging, pulping and manufacturing
processes are expected to conform to the environmental regulations of the
country of origin.

A catalogue record for this book is available from the British Library.

A catalog record for this book is available from the Library of Congress.

10   9   8   7   6   5   4   3   2   1
18   17   16   15   14   13   12   11   10   09

Tranferred to Digital Printing in 2010

Contents

Figures

Preface

Around halfway through my research for this project, I presented some of my findings to a small audience of students and academics at the University of Kent. Immediately afterwards, a debate arose between many of the parties present as to the effectiveness and indeed the verisimilitude of the way in which I had theorised and presented my experiences as a playgoer. In my eagerness to adopt a suitably 'scholarly' tone, I had, argued some, neglected the more phenomenological aspects of theatregoing, attempting to categorise neatly, something which, by its subjective nature, cannot be fully expressed in rational, scientific terms. It was suggested to me by way of illustration that while a psychologist could accurately describe the process of falling in love in a detached and systematic manner, it would never be as full or as complete a description as might be found in a sonnet.

While I am sure my colleague did not want me to write this book in rhyming pentameter, I took her suggestion very seriously. I had long been grappling with ways in which to synthesise my subjective experience (both as a playgoer and as a maker of theatre) with the more formal registers of academic writing; phrases like 'I felt' and 'it seemed' sat ill at ease beside a more strictly analytical mode of writing, and the conclusions I drew were inevitably very personal and utterly unverifiable. I soon realised, however, that my mistake had been to attempt any kind of synthesis at all.

This book, as will become apparent, is concerned throughout with the conflicting registers of performance which often exist unreconciled within the same theatrical event, and with the tension between identification and critical detachment that this instils within its audience. It seems logical that form should echo content, and as such I have woven in between the more conventionally 'scholarly' chapters a thread of unashamedly subjective narratives, largely untheorised, and employing a wider range of writing registers. I am unapologetic that I have made no attempt to make these disparate elements cohere; they will, I hope, bring at least a measure of theatre's plurality of meaning to the book itself.

If not, you can always skip straight to the next scholarly bit.

Acknowledgements

The author and publisher wish to thank the following for permission to reproduce copyright material:

- Artbox/Vestuport for the image from *Romeo and Juliet* (2003);
- Adam Meggido and The Sticking Place (www.thestickingplace.com) for the image from *Shall We Shog?* (2005);
- Garth Hagerman and Daniel Singer for the image of the Reduced Shakespeare Company (1987);
- Steve Tanner and Kneehigh Theatre for the image from *Cymbeline* (2007);
- Dash Arts and The Corner Shop PR for the image from *A Midsummer Night's Dream* (2006);
- The Rose Theatre, Kingston, for the image of the theatre (2007).

Every effort has been made to trace rights holders, but if any have been inadvertently overlooked, the publishers would be pleased to make the necessary arrangements at the first opportunity.

Thanks are also due to the many theatre practitioners who gave up their time to discuss their ideas with me, including John Russell Brown, Annie Castledine, James Garnon, Scott Handy, Richard Henders, Marcello Magni, Malcolm Rennie, Toby Sedgwick, and Pete Talbot. The ideas which form the basis of this book owe much to the innumerable discussions I have had over the years with delegates at the annual British Graduate Shakespeare Conference, Stratford-upon-Avon; I should also like to express my gratitude to all the staff and students of the Drama department at the University of Kent – particularly those who participated in my exploratory workshops, and all those who took part in the various debates about performance documentation which proved so influential in determining the eventual shape and form of this book. I should particularly like to thank Professors Robert Shaughnessy and Russell Jackson for their warm and encouraging responses to the project, and Paula Kennedy and Steven Hall at Palgrave Macmillan for their swift and enthusiastic work in bringing the book to print. I must also thank the Arts and Humanities Research Council, without whose financial support this project would have been very difficult.

I am enormously grateful to the actors (and friends) with whom I have been lucky enough to work over the last few years: Sarah Norton, Martin Gibbons, Caitlin Storey, Mark Hayward, Dominic Conway, Oli Seadon, Dave Hughes, Daisy Orton, Nina Satterley, Clare Beresford, Tom Hughes, Bethan Morgan, Holly Hawkins, Iain Tessier, Ian McKee, Sam Pay, James Groom, and Haydn Pryce-Jenkins. Without their creativity, enthusiasm, imagination, and willingness to experiment, this book would have remained untouched by the many insights generated by our practical work together.

Lastly, I'd like to offer special thanks to Oliver Double, for his guidance and friendship; to Kasia Ladds, for her inexhaustible patience and encouragement; to Anna Purcell, for her keen intelligence and close critical eye; and to my parents, John and Jan Purcell, for their unwavering support.

Personal Narrative 1
Ambiguous Applause

On 1 July 2006, just as the England football team were beginning what was to be their final match in the 2006 World Cup, I found myself a willing participant in what was afterwards a surprising moment of crowd behaviour. I was not at the Veltins Arena in Gelsenkirchen, nor was I (to the amused bewilderment of many of my friends) even watching the match on television. I was at Shakespeare's Globe on London's South Bank, applauding actor Shaun Parkes as he made his final exit as Aaron in *Titus Andronicus*.

This is, perhaps, nothing striking in itself; Parkes's performance had been an arresting and engaging one, making the most of the Globe's capacity for actor–audience interplay with Aaron's frequent asides, and provoking loud and genuine peals of laughter with 400-year-old lines (no mean feat) such as Aaron's exchange with Chiron and Demetrius after their realisation that he has fathered their newborn brother:

DEMETRIUS. Villain, what hast thou done?

AARON. That which thou canst not undo.

CHIRON. Thou hast undone our mother.

AARON. Villain, I have done thy mother.

(4.2.73–6)

His diatribe in 5.1, in which Aaron gloats over his evil deeds and revels in his own callousness, had been a virtuoso delivery. Bare chested, tied to a wheeled platform in the middle of the yard, and surrounded by playgoers, Parkes's gleeful depiction of Aaron's defiant atheism and amorality provided a strangely attractive contrast to David Sturzaker's cold and

1

hypocritical Lucius, who stood in judgement over him upon the stage. Lucius has, by this point in the play, been party to many of the actions in the play's bloody cycle of violence, and his claims to moral authority are therefore highly suspect. In Lucy Bailey's staging, Parkes's Aaron did of course have the advantage of a spatial alignment with the audience in this scene – he, like us, was standing in the yard, looking up at the stage – but there was something more. Aaron's unashamed villainy in the face of certain death seemed somehow infinitely more honest than Lucius's strategic detachment, a detachment which moments earlier had enabled him to threaten Aaron's baby son with hanging ('First hang the child, that he may see it sprawl', 5.1.51). Lucius seemed to stand, at this moment in the play at least, as a representative of authority's moral ambivalence in the face of war and bloodshed (an ambivalence surely not far from audience's minds in the context of the ongoing conflicts in the Middle East), and Aaron's invective, despite (or perhaps partly because of) its barefaced depravity, became a wild and emphatic two-fingers-up to Lucius and all he represented. It was unsettling, certainly – when Aaron mentioned his desire to 'ravish a maid', Parkes looked hungrily and menacingly at a nearby female audience member – but it instilled in the crowd a distinctively subversive energy. This energy, it seemed to me, almost threatened to spill over after the play's bloody denouement as Marcus stood beside Lucius (on the very same platform to which Aaron had earlier been tied) and asked the surrounding playgo-ers, addressing us as his compatriots, 'What say you, Romans? / Have we done aught amiss?' (5.3.127–8). I, for one, could feel the response 'Yes!' forming on my lips.

When, following this, Aaron was brought back into the yard and sentenced by Lucius to be 'fastened in the earth' and left to starve, the short, seven-line speech he gave as he was dragged away to his death recalled all the subversive energy of his previous scene. He roared the final lines through the crowd as he left:

> Ah, why should wrath be mute and fury dumb?
> I am no baby, I, that with base prayers
> I should repent the evils I have done.
> Ten thousand worse than ever yet I did
> Would I perform if I might have my will.
> If one good deed in all my life I did
> I do repent it from my very soul.
>
> (5.3.183–9)

The applause was natural and unprompted, a collective and spontaneous gesture of approval, and it was only afterwards, as I headed towards the Founders Arms to watch England lose, once again, on penalties, that I realised I had been applauding a sentiment which was distinctly immoral. I comforted myself that it was Parkes's energetic performance which had drawn my, and the rest of the audience's, applause; but deep down, I knew it had not been quite so clear-cut. What had I applauded? What had I endorsed? What had I, along with the hundreds of other people in the Globe that day, expressed in such a public and communal manner?

It continued to bother me as a spot kick from Cristiano Ronaldo sealed the match in Portugal's favour, consigning England's football fan base to another four 'years of hurt', and sending the crowd gathered in the Founders Arms off into London's early evening with a few disappointed shrugs and a surge of collective indifference.

1
Popular Shakespeares

In September 2007, a theatre company called The Factory embarked on a series of performances titled *The Hamlet Project*. They had been preparing for this project for several months: primarily under the supervision of director Tim Carroll, but in collaboration with a wide variety of practitioners from across the contemporary British theatre. The fundamental principle of the project was, and is, that the same show will never be performed twice.

The Factory have performed *Hamlet* nearly every Sunday since the project began, in a different venue every week (generally in and around London). So far, these have ranged from theatrical studios and converted churches, to a Victorian pumping station, to the Debating Chamber of London's County Hall. Each show has a different cast: actors, having learnt a variety of roles, play the chance game Paper-Scissors-Stone moments before the show starts, to determine who will play the central roles. 'Obstructions' are given to the performers at the beginning of every act – they might, for example, have to ensure that they are always standing in a different part of the space from the character to whom they are speaking, or that they never depart more than a metre from a given spot. A key ingredient of the production's distinctly irreverent character is its insistence that the only props used in any performance will be those brought along by the audience. Audiences – ever keen to break with theatrical convention when challenged to do so – will bring along anything from confectionery to sex toys.

I open this book with an overview of this project because it brings together in one show much of what is being done across innumerable forms of Shakespearean performance in Britain today. The actors of The Factory adhere very strictly to the metre of the text, but through their incorporation of the conspicuously extra-textual props donated by the audience, they simultaneously (and contradictorily) undercut its Shakespearean authority. The show engenders a distinctly 'unofficial' feel, attracting a predominantly young audience almost entirely through word of mouth, email, and social networking sites (the production has no publicity budget whatsoever). It maintains a dual focus which is at once both inside and outside the world of the play: we may be watching a story about the Prince of Denmark, but we are also watching actors negotiate a set of real-world performance challenges. Conventional divisions between actors and audience are constantly subverted – performances take place in shared light among and around audience members, with no space stably 'on-' or 'off-stage'. The audience's collaborative role in the creation of the performance is foregrounded, as playgoers are not only addressed directly but frequently involved physically in the action. The production draws attention to its own 'liveness', emphasising the unrepeatability of its moment-by-moment theatrical discoveries, and often flirting with actual physical danger (in a performance I saw, one actor delivered a speech while balancing precariously on a wall on the very edge of the River Thames). In a show at the Southwark Playhouse, the actor playing Hamlet picked up a real baby in order to illustrate 'What a piece of work is a man' (2.2.305–12). The *Evening Standard* described the moment as 'heart-in-the-mouth theatre, unbearably poignant, unrepeatable – screw health and safety' (17 January 2008).

Many of these trends and concerns are ones which crop up again and again throughout contemporary Shakespearean performance, and they are characteristic of the cultural phenomenon which is the subject of this book. *The Hamlet Project* did not appear from nowhere. As I hope this book will show, it grew from, and in, a very specific, if bewilderingly multifaceted, cultural moment. The phenomenon which I would like to call 'Popular Shakespeare' is at once ubiquitous and elusive, spanning fringe productions, mainstream theatre, the mass media, and indeed culture at large.[1] Not just a radical alternative to high-culture Shakespeare, it represents a shift, or perhaps to put it more precisely, an interrelated assortment of shifts, in what the name 'Shakespeare' means to us today.

Immediately we hit upon the problem of scope. There is no neat line one can draw around such a subject, separating it from other areas of culture; Shakespearean theatre does not exist in a vacuum, but is part of a spectrum of related and interconnecting cultural arenas, from stand-up comedy and sitcom to advertising, blockbuster films, and television sci-fi. As a result, then, although on one level this book provides an account of Shakespearean performance in Britain since the 1990s, its subject – 'Popular Shakespeare' – is far more wide-ranging. I do not confine my discussions to the British, the live, or even exclusively the contemporary; as the reader shall see, such categories are hardly stable ones in any case.

As the Personal Narrative strand suggests, this book makes no attempt to deny its author's subjectivity. This does, I hope, have its advantages as well as its limitations. My own theatregoing, and indeed my cultural experience more broadly, is overwhelmingly UK-based: the reader will find, therefore, that the 'Popular Shakespeares' discussed here are on the whole British ones. This is ultimately a study of one observer's encounters with a set of cultural phenomena, and another writer in a different cultural context might experience the same subject very differently. (I have certainly found as much, for example, in my own readings of accounts by American commentators of spectatorship at the reconstructed Shakespeare's Globe. When W. B. Worthen writes that the 'performativity' of Globe audiences is frequently shaped 'by expectations, modes of attention, and habits of participation' learned from 'living history' theme parks and historical re-enactments (2003: 86), I wonder whether his analysis is truer of American audiences than it is of British ones; living history museums and battle re-enactments are, I believe, more widely experienced in the US than they are in the UK.) In Britain, 'Shakespeare' means and embodies a set of values which are very different from its cultural meanings elsewhere in the world (though I do not intend to suggest that these values are in any sense fixed or coherent). I hope, therefore, that this study's use of its own author as an eyewitness will allow it to root its discussions in as specific a cultural context as possible.

This is not to suggest, however, that I intend to dispense entirely with discussion of non-British Shakespeares: far from it. Indeed, it is often productions visiting the UK from countries where Shakespeare is not such a revered symbol of national identity and cultural authority that provide the most iconoclastic challenges to British cultural norms. The reactions to such productions by the theatre critics of the British press – often, as we shall see, our self-appointed cultural guardians – are a major part of this study. The book will also make frequent reference to American productions, both live and recorded. In doing so, I do not wish to suggest that the trend for 'Popular Shakespeare' is taking place in identical terms on either side of the Atlantic; certainly 'Shakespeare' is subject to dynamic cultural contestation in the US just as it is in the UK, but a study of the debates, histories, and ideologies with which the name is bound in America would be a study unto itself.[2] Rather, I consider the contribution of American constructs of Shakespeare towards the shifting cultural status of Shakespeare in Britain. British Shakespeare owes much of its hybrid and self-contradictory identity to imported US Shakespeare, particularly due to the proliferation of American cultural products through the mass media. Ultimately, of course, there is no such thing as a distinctly 'British' Shakespeare anyway. Abundant examples of international casts and multicultural influences at such emphatically 'British' theatres as Shakespeare's Globe and the Royal Shakespeare Theatre prove that much like the perennially composite British nation itself, 'British Shakespeare' is always already culturally hybrid.

Investigations into Shakespeare and pop culture are emerging at the forefront of contemporary Shakespearean scholarship (see, for example, Burt 2002; Burt & Boose 2003; Lanier 2002; and Shaughnessy 2007), but an area of this field which remains as yet relatively unexplored is that of Shakespeare and 'the popular' in live performance. This, then, is the central focus of my book: these studies are rooted in performance theory and practice, owing as much to Bertolt Brecht, Peter Brook, and their academic and theatrical descendants, as they do to the work of media and cultural theorists. However, just as one cannot cleanly delineate a 'British' Shakespeare from the plethora of cultural alternatives, one cannot comfortably distinguish the 'live' from the mediatised either. As Philip Auslander has argued, the two categories coexist within one piece of work more often than is commonly assumed, and our experience of the live may ultimately be a 'mediatised' one anyway (Auslander 1999). It is, as we shall see, very frequently the mass media which shapes, echoes, and embodies 'popular' attitudes towards Shakespeare; television and film are so entrenched in the very fabric of popular culture that contemporary popular theatre cannot be studied in isolation from them. Thus, though its focus is on live performance, the book will also draw examples and analysis from the mass media.

The book looks primarily at contemporary performance, which here is taken to mean largely post-1990 (though naturally one cannot entirely separate the 'contemporary' from its immediate forebears, and consequently some examples are drawn from earlier in the twentieth century). It attempts to chart – though inevitably only partially and incompletely – something of Shakespeare's changing cultural status in the wake of factors such as the theatrical innovations of the 1960s and beyond, the proliferation of the mass media, the rise of the 'MTV generation', the Shakespeare film phenomenon of the 1990s, and the reconstruction of the Globe. I consider various debates, both in the press and in academia, surrounding Shakespeare and the 'right' way to do it: there emerges a picture of a cultural war between those intent on 'popularising' Shakespeare (for whatever purpose – as we shall see, this varies considerably) and those who are resistant to what may be conceived as a form of cultural contamination. Central to these debates, of course, are notions of authenticity, continuity, cultural heritage, and Shakespearean authority. For this reason, then, though the subject of this book is *contemporary* Shakespearean performance, Chapters 1–3, and then Chapter 6, will root their discussions in some historical and textual analysis. I have done so not to propose that an engagement with popular theatre forms is the 'right' way to stage the texts, but rather to point out that constructions of Shakespeare as inherently cohesive (in the mould of naturalism) and opposed to dialogic performance modes are easily contested, and that modern evocations of a pure, 'literary' Shakespeare have no provable relation to the historical precedent to whose authority they generally (though by no means always) refer.

The term 'popular' itself is the subject of the rest of this chapter, particularly with regard to its use in relation to Shakespearean performance. As we shall see, 'popular theatre' can perhaps best be understood not as a hard-and-fast category of theatre but rather as a way of looking, as an attitude taken up towards a piece among its audience.[3] Even this attitude, however, has been interpreted in wildly differing ways. This chapter, then, will explore the implications of the term, before going on to set up a theoretical model through which we can begin to examine the complicated and sometimes mutually contradictory relationship between 'popular theatre' and Shakespeare.

Conflicting popular theatres #1: The Commercial and the Anthropological

'Popular theatre' is one of those terms which can be easily taken for granted. What it describes would seem to be self-evident: theatre which is 'popular'. Try to define what constitutes 'popular', however, and things start to come unstuck. That the same term might just as easily be applied to *Starlight Express* as to storytelling, to Bernard Manning as to Dario Fo, to mummers' plays as to bedroom farces, indicates that the label implies no shared political standpoint or stylistic features, no distinctive audience demographic, nor any particular measure of commercial success. That even the Royal Shakespeare Company has been self-characterised by two consecutive Artistic Directors as a producer of 'popular theatre' begs the question as to which theatrical forms might *not* be considered part of the genre, and indeed whether the term is so all-encompassing as to be rendered useless for the purposes of scholarly discussion.

The literature on the subject is as inconsistent and as contradictory as the field itself. The radical director and playwright John McGrath used the term interchangeably with 'working-class entertainment' throughout his famous book on the subject, *A Good Night Out*; in *Shakespeare and the Popular Dramatic Tradition*, S. L. Bethell defined the form according to what he called its 'dominant characteristic', 'the audience's ability to respond simultaneously and unconsciously on more than one plane of attention at the same time' (1977: 29). For Joel Schechter, the term is associated 'with democratic, proletarian, and politically progressive theatre' (Schechter 2003: 3); for David Mayer, on the other hand, popular theatre 'beguiles and amuses, rests and relaxes the mind, encourages conviviality and satisfaction with things as they are' (1977: 266). Whereas Peter Brook has argued that 'the popular theatre is anti-authoritarian, anti-traditional, anti-pomp, anti-pretence' (1990: 76), Mayer suggests that popular drama is likely to be 'of a traditional nature' and that 'sensational scenes' and amusing character types are likely to be 'favoured over situations that deal with immediate moral and social values in a meaningful way' (1977: 264–5).

The root of the term's fuzziness lies, I would suggest, in the various possible meanings of the word 'popular', many of which are fundamentally at odds with one another. The sixth entry in the *Oxford English Dictionary* definition gives 'popular' as 'Finding favour with or approved by the people; like, beloved, or admired by the people, or by people generally; favourite, acceptable, pleasing.' This is what Stuart Hall describes as the 'market' definition of the word, where things are said to be popular 'because masses of people listen to them, buy them, read them, consume them, and seem to enjoy them to the full'. He notes that this form of popularity is 'quite rightly associated with the manipulation and debasement of the culture of the people' (1998: 446). In this sense, the West End musical might be said to be the most 'popular' of Britain's theatrical forms; produced with the principal aim of turning a profit for its investors, it relies on tried-and-tested theatrical effects in order to attract as many customers as possible, and fits in with Mayer's model of a popular theatre which 'encourages conviviality and satisfaction with things as they are' (1977: 266).

In this sense, there might be little or no difference between the 'popular' and the kind of 'populist' theatre discussed by Peter Holland in *English Shakespeares.* Holland suggests that 'popularised' productions such as the RSC's 1990 *Comedy of Errors* 'seemed generated more by a capitulation to the RSC's perennial funding crisis than by a wish to see what the play could offer' (1997: 62), and argues that the English Shakespeare Company's attempts at 'popular' Shakespeare in their 1991 productions of *The Merchant of Venice* and *Coriolanus* 'made the productions weakly populist, offering simple answers where the text is complex' (1997: 95).

In equating 'popular theatre' with populism, however, I wonder whether Holland is conflating two very different ideas. I would suggest that we might usefully make a distinction between theatre which is commercially 'popular' ('populist') and the term 'popular theatre', since the latter's theatrical etymology is very different from the most common usage of the word. Derived from the French *théâtre populaire* – according to Schechter, Jean-Jacques Rousseau used the term in 1758 (2003: 3) – 'popular' in this sense stems more directly from its Latin ancestor *populus* ('the people'), implying a theatre of the people, speaking to them in their own idioms, voicing their own concerns, representing their own interests. Hall describes this understanding of the word as 'anthropological'.

It was presumably this model of popular theatre, or at least something like it, which Peter Hall had in mind when he stated the RSC's ambitions in 1963:

> We want to run a popular theatre. We don't want to be an institution supported by middle-class expense accounts. We want to be socially as well as artistically open. We want to get people who have never been to the theatre – and particularly the young – to see our plays.
>
> (Addenbrooke 1974: 63)[4]

The mainstream theatre audience was, as Hall implies, predominantly middle class, middle-aged, and well educated; it is a characterisation that holds as true now as it did in 1963. Presumably the first qualifying factor as a piece of popular theatre in the anthropological vein is, therefore, an attempt at social extension; a sense that the piece is concerned, as Mayer puts it, 'with the widest reach of audience available at a given moment or place' and that it is offered for 'the enjoyment or edification of the largest combinations of groupings possible' within a given society (1977: 263). A typical popular theatre audience, Mayer suggests, might consist of 'farmers, artisans, factory workers, shopkeepers, labourers, the rural and urban poor and middle classes' (1977: 263). By this reasoning, then, popular theatre is defined as such by its *audience* and not by its form at all. Thus Joan Littlewood and the Theatre Workshop's attempt at creating a truly popular theatre in a working-class area of London's East End was frequently thwarted by the copious and conspicuous presence of wealthier playgoers from other parts of the city; and by the same logic, though in its outward form it was unmistakeably a part of the 'popular theatre' tradition, the Old Vic's 2005 pantomime of *Aladdin* starring Sir Ian McKellen as Widow Twankey priced itself out of the genre by making itself affordable for the most part only to families with relatively high incomes.

Here, popular theatre is defined in its distinction from theatre which is socially exclusive; or rather as Joe Farrell argues, the opposite of popular theatre is not theatre which is 'unpopular' in a commercial sense, but theatre which is *bourgeois* (1991: 3). At this point, the West End musical, with its high ticket prices and implied intimations of social aspiration, starts to disappear from the picture. But the real difficulty in distinguishing the 'popular' from the 'bourgeois' arises from the fluidity of both categories. Those forms which appeal to broad sections of the population may, as we have seen, serve only to reinforce the hegemony of the bourgeoisie; and in any case, as Stuart Hall points out, 'from period to period, the contents of each category change. Popular forms become enhanced in cultural value, go up the cultural escalator – and find themselves on the opposite side' (1998: 448). As Baz Kershaw has argued, partly in response to McGrath's simplistic model of 'working-class entertainment',

> [i]t is a mistake to suggest, for example, that forms such as pantomime and music hall once *belonged* to the working classes and have been appropriated subsequently by enemies of the class. Those forms almost always were part of a complex dialectic through which conflicting ideologies, conflicting interests, have been staged.
>
> (1992: 154)

As Kershaw suggests, popular theatre must be considered not only in relation to the audience to which it appeals, but also with respect to what it *does*, sociologically and ideologically. While as we shall see, all forms of popular theatre are complicated by an element of political ambiguity, there must be a sense that any form of 'people's theatre' is indeed serving the interests of 'the people'. As playwright David Edgar argues in *State of Play*, 'without a provocative agenda somewhere in the vocabulary,' the 'popular' inevitably degrades 'into the plebeian and the philistine' (Edgar 1999:17).

It was this understanding of 'popular' which lay at the heart of Bertolt Brecht's use of the term. In his 'fighting conception' of the word,

> '[p]opular' means intelligible to the broad masses, taking over their forms of expression and enriching them / adopting and consolidating their standpoint / representing the most progressive section of the people in such a way that it can take over leadership: thus intelligible to other sections too / linking with tradition and carrying it further / handing on the achievements of the section now leading to the section of the people that is struggling for the lead.
>
> (Brecht 1977: 108)

Such a 'popular' theatre was presumably what Trevor Nunn had in mind for the RSC when he spoke in 1974 of his desire for 'an avowed and committed popular theatre': 'I want a socially concerned theatre. A politically aware theatre' (Addenbrooke 1974: 182).

Precisely what a 'socially concerned' theatre should do to its audience is, however, an open question. 'Socially concerned' theatre critics and practitioners are, as we shall see, divided on the matter; and where the subject under discussion is not leftist agitprop, but Shakespeare (itself a politically ambiguous subject), the issue naturally becomes further complicated. When I began this research project, my understanding of 'popular theatre' was that it should build and strengthen the temporary community within its audience in order to be classified as such; an argument along these lines is laid out in my article 'A Shared Experience: Shakespeare and popular theatre' which appeared in *Performance Research* in 2005. The following section, I hope, shows the matter to be rather more complex.

Conflicting popular theatres #2: The Collectivist and the Critical

There are certain forms which have historically been associated with popular theatre; forms which Theodore Shank describes as being associated 'with fun, with the common man, with people in general regardless of social, economic, or educational status'. He gives the following examples, though the list could easily be ten times longer:

> [C]ommedia dell'arte, circus, puppet shows, music hall, vaudeville, parades, magicians, carnival side-shows, buskers, brass bands, comic

strips, striptease, melodrama, minstrel shows, and other means of exhilarating celebration.

(2003: 258)

'Through the ages it has taken many forms', says Peter Brook in his chapter on popular theatre in *The Empty Space*, 'and there is only one factor that they all have in common – a roughness' (1990: 73).

It is this 'roughness' – that Mike Alfreds, founder of the theatre company Shared Experience, has described as 'the "rough" quality implicit in the unknown inter-action between actors and audience from night to night' (1979: 4) – which underpins much of the theory behind this book. Virtually every popular theatre form draws attention to the concrete elements of its own production: masks, puppets, audience interaction, spectacle, technique, physical risk – fundamentally, as McGrath puts it, a 'non-elaborateness of presentation' (1996: 28). As both Brook and McGrath argue, popular audiences are used to switching between modes of reception, 'accepting inconsistencies of accent and dress, ... darting between mime and dialogue, reason and suggestion' (Brook 1990: 75) – in other words, following a story without ever losing awareness of the means by which it is being told. Such 'rough' theatre, explains Alfreds,

> creates what isn't there at all through the creative will of the actor and the imaginative complicity of the audience. One reveals the nakedness of one's techniques, one hides nothing and the results are usually more convincing than the most elaborate lighting and scenery.
>
> (1979: 10–11)

Actors and their audience share a collective imaginative experience, and – so the theory goes – are 'brought together by it' (Alfreds 1979: 4). This breed of 'shared experience' is designed to prompt a kind of communality or even 'communion' among its spectators, which might be profitably explored with reference to anthropologist Victor Turner's concept of *communitas*.

Communitas describes the intense feelings of solidarity and togetherness which can be instilled among members of a large group of people, be it in a crowd at a football stadium, at a rock concert, at a religious ceremony, at a political rally, or, of course, at a theatrical event. Turner explains it as a moment of liminality in which the group somehow steps outside of its normal social parameters, and social models alternative to the established order become equally possible. It is, says Turner, 'a direct, immediate and total confrontation of human identities' (1969: 132):

> Spontaneous communitas has something 'magical' about it. Subjectively there is in it the feeling of endless power. ... It is almost everywhere held to be sacred or 'holy,' possibly because it transgresses or dissolves the norms that govern structured and institutionalised relationships and is accompanied by experiences of unprecedented potency.
>
> (1969: 128–39)

Certainly there are echoes here of Carnival, in which, as Mikhail Bakhtin describes,

> [t]he individual feels that he is an indissoluble part of the collectivity, a member of the people's mass body. ... Carnival with all its images, indecencies, and curses affirms the people's immortal, indestructible character. In the world of carnival the awareness of the people's immortality is combined with the realisation that established authority and truth are relative.
>
> (1984: 255–6)

It was perhaps the same sense of togetherness which the former Artistic Director of Shakespeare's Globe, Mark Rylance, described in a talk at the Globe in 2005, when he stated his belief that 'People in a group have a collective imagination that is larger than any one of us individually. ... Collectively we are able to perceive more things than we are on our own.'[5] His words, I felt, were strongly reminiscent of Peter Brook's definition of the 'way of theatre' as 'leading out of loneliness to a perception that is heightened because it is shared' (1989: 41), and indicative of an attitude prevalent among many theatre practitioners concerned with popular theatre.

Such communion is not, however, politically unproblematic. In his 1997 review of the Globe's *Henry V*, Paul Taylor identified the Globe's 'amazing sense of audience solidarity' as a strength, but wondered whether director Richard Olivier had done enough to put the 'atavistic jingoism' of the scenes in which the 'dastardly French' characters were booed into perspective (*Independent*, 7 June 1997). The strengthening of a group identity serves to reify that group's norms, and if those norms are conservative, then so is the theatrical experience.

In *The Audience*, Herbert Blau questions the very idea of 'the audience as community, similarly enlightened, unified in belief, all the disparities in some way healed by the experience of theater' (1990: 10). Blau sees this as 'a notion of community that could never be satisfied in the theater as anything but a fiction', and suggests that it is a dangerous fiction at that; he uses the term 'participation mystique' to refer to the phenomenon (1990: 11). Coined by French ethnologist Lucien Lévy-Bruhl and made famous by Jung, this term describes a state of mind in which no differentiation is made between the self and things outside the self. Unthinking engagement with a group consciousness of this sort can lead to a mob mentality, in which all capacity for independent, critical thought is lost; historical horrors such as lynchings or the Nuremberg rallies are attributable to it. As Blau puts it:

> [W]e have had only too woeful evidence in the twentieth century that the most complex societies can be profoundly, hysterically, and devastatingly moved.
>
> (1990: 18)

Blau argues instead for Brecht's 'willingness to forgo the blessed moment of intimacy for the rigorous moment of perception' (1990: 5–6); we shall explore Brecht's position on the matter in a moment.

I am not sure whether *communitas* is quite the dangerous fiction that Blau suggests, however, and would suggest that in much popular performance it is closer to *consensus* than it is to participation mystique. Theatre practitioner Bim Mason acknowledges the potentially negative effects of the 'diminishing of the ego by merging into the same rhythm as a large group' by likening it to the 'loss of individualism' encouraged in modern armies by the parade-ground drill (1992: 39). He points out, though, that on the other hand such theatre can achieve political efficacy by giving the community concerned 'a positive sense of solidarity and power' (1992: 40). As Turner argues, the power of *communitas* 'is no substitute for lucid thought and sustained will' (1969: 138); but as John McGrath has suggested, theatre 'can be the way people can find their voice, their solidarity and their collective determination' (1974: xxvii). Chapter 5 will explore this problem in more detail.

If we have on the one hand, then, a model of what we might call a 'collectivist' popular theatre seeking to affirm and consolidate a communal identity, we have on the other a popular theatre which seeks to disrupt and destabilise that same identity in order to instil a 'critical' attitude among that community's individual members. A third model, however, combines the two. While Kershaw states that 'collective responses, shaped by the ideological identity of the audiences' communities, are the very foundation of performance efficacy' (1992: 35), he argues that if a performance is to achieve political efficacy then it must both confirm *and* subvert the communal values of its audience:

> [T]he company must reassure the community that it is able to represent its interests. In contrast to this, and in order to approach efficacy, the performing itself must employ authenticating conventions/signs to discomfort or disturb the ideology of the community.
>
> (1992: 32)

The effect is, to use a well-worn theatrical quotation, 'to make the spectator adopt an attitude of inquiry and criticism' towards his or her own community values (Brecht 1977: 136).

And so we come, inevitably, to Brecht. As we have seen, critics such as Blau have read Brecht in opposition to what I have termed the 'collectivist' theatre; certainly Brecht himself condemns the notion of a unified audience, making it quite clear that the actor does not treat the audience as 'an undifferentiated mass':

> He doesn't boil it down to a shapeless dumpling in the stockpot of the emotions. He does not address himself to everybody alike; he allows the existing divisions within the audience to continue, in fact he widens them. He has friends and enemies in the audience; he is friendly to the one group and hostile to the other.
>
> (1977: 143)

But what Brecht does imply is a split in the audience's consciousness. Far from the common misconception of Brechtian theatre as having 'no interest in generating audience emotion' ('dooming those of us on the left to a theatre without thrills', complains Simon Shepherd, 2000: 224–5), Brecht in fact states quite clearly that 'Neither the public nor the actor must be stopped from taking part emotionally' (1965: 57). One of his appendices to *A Short Organum for the Theatre* makes this clear:

> The contradiction between acting (demonstration) and experience (empathy) often leads the uninstructed to suppose that only one or the other can be manifest in the work of the actor. ... In reality it is a matter of two mutually hostile processes which fuse in the actor's work; his performance is not just composed of a bit of the one and

a bit of the other. His particular effectiveness comes from the tussle and tension of the two opposites, and also from their depth.

(Brecht 1977: 277–8)

As Shomit Mitter has argued in *Systems of Rehearsal*, alienation *relies* upon an element of emotional involvement (which by its nature is communally experienced)[6] since it is effective 'only in proportion to the emotional charge it undercuts' (Mitter 1992: 49). Mitter's chapter on Brecht is, rather aptly, entitled 'To Be And Not To Be', since in it he argues that Brechtian actors

> must not merely learn to be and not be their characters; they must at all times be able to move uninhibitedly between these positions – for defamiliarisation is not a position but the result of a transition.
>
> (Mitter 1992: 55)

Here we see that Brechtian theatre, unlike the single-level dogmatism of much agitprop, is in fact all about maintaining a plurality of meaning, and not so much didactic as *dialectic*. Here, the similarity to Bethell's understanding of the popular audience's ability 'to respond simultaneously and unconsciously on more than one plane of attention at the same time' – what he calls 'the principle of multiple consciousness' – becomes clear (1977: 29).

Bakhtin's theory of the carnivalesque, in fact, also hinges upon a pluralism of this kind. Talking of the power of shared laughter – which he describes as both 'ambivalent and universal' – Bakhtin extols its ability to liberate 'from the single meaning, the single level' (1984: 122–3). David Edgar describes 'this capacity to express the opposite in the same plane' as, for him, 'the most exciting aspect of carnival as Bakhtin defines it' (1988: 242–3), and it is this capacity which differentiates the 'indissoluble collectivity' of Bakhtin's carnival crowd from the 'undifferentiated mass' condemned by Brecht. A combination of *communitas* and critical distance might, as Edgar suggests, be possible:

> What I suppose most of us are striving for, is a way of combining the cerebral, unearthly detachment of Brecht's theory with the all too earthy, sensual, visceral experience of Bakhtin's carnival, so that in alliance these two forces can finally defeat the puppeteers and manipulators of the spectacle.
>
> (1988: 245)

Edgar here proposes a popular theatre which bestrides two very different planes of experience, switching between *communitas* and independent, critical thought, at once uniting and deconstructing, alternately affirming and then disrupting or destabilising group identities. If any theatrical form is capable of such pluralism, it is (as both McGrath and Brook have suggested) the popular theatre.

To conclude, then, we have looked at two sets of conflicting models of popular theatre: the 'Commercial' versus the 'Anthropological', and the 'Collectivist' versus the 'Critical'. As we have seen, however, the fact that these models are contradictory does not mean they cannot exist, unreconciled, within the same theatrical event. A piece might be commercially popular at the same time as appealing to a wide and varied audience. It might reinforce some social norms while disrupting others; the attitude it instils among its audience might be both celebratory and critical, confirming and disrupting its group identities. An example of a theatre falling into all four categories might be the reconstructed Shakespeare's Globe. As Douglas Lanier points out, this theatre's dependence upon the academy, the heritage industry, tourism, and sponsorship from international businesses compromise 'its claim to present a people's Shakespeare' (2002: 165–6); at the same time, its inexpensive ticket prices (600 are available for £5 for each performance) and widespread appeal ensure that audiences are socially and culturally diverse. In the theatre's 2005 production of *Pericles*, eye-catching stunts and physical comedy provoked waves of communal responses throughout the show, and the audience started from time to time to adopt patterns of group behaviour more associated with sport than with theatre. In the same production, however, Patrice Naiambana's Gower, interacting with the audience throughout in the manner of a stand-up comic, was unafraid to disrupt the audience's feelings of togetherness, goading them about their resistance to multicultural and heavily adapted performances of Shakespeare and playfully suggesting an inherent racism. The production may have been politically ambivalent, but it was all the more interesting for it.

Shakespeare as popular theatre #1: A textual approach

In his groundbreaking study *Shakespeare and the Popular Tradition in the Theater*, Robert Weimann too identifies an 'extremely effective balance between dramatic enchantment and disenchantment' (1987: 247) as a key feature of popular theatre. It is one shared, he argues, with Elizabethan drama, in which he finds evidence for a 'dual perspective' encompassing 'conflicting views of experience' (1987: 243).[7]

Certainly Weimann is not alone in this analysis. Bakhtin maintained that the most important works in which 'the two aspects, seriousness and laughter, coexist and reflect each other' are Shakespeare's tragedies (1984: 122), while Brecht himself described Shakespeare's theatre as 'a theatre full of A-effects' (1965: 58). Even in 1632, the author of the second folio's commendatory poem *On Worthy Master Shakespeare and His Poems* noted such an effect:

> to temper passion that our ears
> Take pleasure in their pain, and eyes in tears
> Both weep and smile: fearful at plots so sad,
> Then laughing at our fear; abused, and glad
> To be abused, *affected with that truth*
> *Which we perceive is false*; pleased in that ruth
> At which we start, and by elaborate play
> Tortured and tickled; ...
> This, and much more which cannot be expressed
> But by himself, his tongue and his own breast,
> Was Shakespeare's freehold.

(ll. 23–42; my emphasis)

Weimann's key illumination was to analyse this phenomenon in terms of *locus* and *platea*, borrowing his terminology from early morality drama: in the pageant theatre, the *locus* was the localised, representational area upon the platform, while the *platea* lay in the street below, the 'broad and general acting area in which the communal festivities were conducted' (1987: 79).

The spatial functioning of *locus* and *platea* in its purest form worked something like this: the actor standing upon the scaffold would appear against the backdrop of the pageant stage – the throne, for example – which would localise him in a specific, or at least relatively specific, fictional setting. The actor standing in the *platea* space, however, would have been standing against no such backdrop; behind the actor, the majority of the audience would have been able to see only other members of the audience, and the performance became, by implication, an extension of their own world.

With the *locus*, Weimann identifies 'a rudimentary element of verisimilitude' (1987: 75); characters would be presented without direct reference to the world of the audience, and would often be of a high social status. As such, the *locus* was the site of heightened language, elevated subject matters, and officially sanctioned historical narratives. The *platea*, on the other hand, as 'a theatrical dimension of the real world' (1987: 76), would be the site of direct address and audience interaction; low status characters such as rustics and clowns would use the vernacular language and provide a profane, satirical, and subversive (and often anachronistic) counter-perspective to the affairs of the *locus*.

As applied to an analysis of the Elizabethan theatre, however, the matter becomes less clear-cut. Rather than using *locus* and *platea* and their associated modes of performance in distinct contrast to one another, Weimann suggests, it was the areas *between* these two poles which were explored by Shakespeare and his contemporaries in their drama. 'What is involved,' he argues,

> is not the *confrontation* of the world and time of the play with that of the audience, or any serious *opposition* between representational and non-representational standards of acting, but the most intense *interplay* of both.
>
> (1987: 80)

Some of Shakespeare's most powerful moments actively encourage movement in and out of the illusion – Macbeth's metatheatrical speech about the 'poor player, / That struts and frets his hour upon the stage, / And then is heard no more', for example. But *locus/platea* interplay serves a political function, too. According to Weimann, the 'plebeian intermediaries' who occupy a *platea*-like position in Shakespeare's drama 'help point out that the ideas and values held by the main characters are relative to their particular position in the play' (1987: 228); Phyllis Rackin has argued that by its very nature, the anachronism of the *platea*-like characters points to the constructedness of the official histories of the main plot (1991: 104). Thus the *platea* register serves as a means of destabilising the discourses of the *locus*.

The Nurse from *Romeo and Juliet* might provide a quick example. As a character who interacts regularly with characters from both registers of performance (Peter and the clowns on the one hand, Juliet and Romeo on the other), the Nurse occupies a liminal space between *platea* and *locus*. Her presence can function in performance as a subversive counter-perspective to the earnestness of the central plot: she jokes with the audience about Paris's sexual insatiability just before she discovers Juliet's supposed corpse in 4.4, and even debases the scene in which Juliet's parents bemoan their loss with her own overblown lamentation:

O woe! O woeful, woeful, woeful day!
Most lamentable day! Most woeful day
That ever, ever, I did yet behold!
O day, O day, O day, O hateful day,
Never was seen so black a day as this!
O woeful day, O woeful day!

(4.4.80–5)

The Nurse's identification with the Bakhtinian 'grotesque body' (with its emphasis on the parts of the body through which things enter and exit) is evident from the outset: in only her first scene, she makes reference to her rotten teeth, discusses breastfeeding at length, and jokes about Juliet's future sexual awakening and pregnancy.[8] Her disarmingly cynical and overtly sexual attitude towards love provides a satirical counter-perspective to Romeo and Juliet's impulsive and passionate affair throughout the play.

As You Like It provides a more complex example. During the course of the play, the boy-actor playing Rosalind has adopted not only the guise of a woman, but of woman-playing-boy (Rosalind as Ganymede) and of woman-playing-boy-playing woman (Rosalind-as-Ganymede as 'Rosalind' in his/her game with Orlando). Naturally, none of these roles are stable, but rather subject to metatheatrical gender jokes throughout the play which disrupt the coherence of each. By the time the actor returns to the stage at the end to present the epilogue, all but one of these roles have been stripped away; but in a moment of gender-bending liminality, the actor plays both male and female at once. 'It is not the fashion to see the lady the epilogue,' the actor begins (1–2), presumably still in the female (*locus*) performance register. But within moments, the line 'If I were a woman, I would kiss as many of you as had beards that pleased me, complexions that liked me, and breaths that I defied not' (16–19) implies that we are now hearing the words of the actor as a male (*platea*). As he/she makes his/her final curtsy, the audience is almost inevitably left with a destabilised perception of the actor/character's gender.

Harry Berger illustrates what he calls these 'modal shifts between collaboration and illusionistic emphases' as a 'polar continuum'. He imagines a spectrum with 'actor' at one end and 'character' at the other; in the centre is the nebulous 'character/actor' ('by addressing the theatre audience,' Berger explains, 'the character presents herself or himself as the actor playing that character'; 1998: 49).[9] However, while the idea of a continuum is useful, I am not sure that the interplay between actor and character is as straightforward as Berger's notation suggests. A good actor can encompass both ends of the spectrum in a single moment with the right combination of body language and audience eye contact, and in any case, the extent and nature of such a transition will always be ambiguous. In his recent book *Author's Pen and Actor's Voice*, Weimann analyses a passage of Lance's from *The Two Gentlemen of Verona* as 'not so much a progress, a gradation or change through a series of stages from the definitely localised to the unlocalised, but a recurring, more immediate overlapping or oscillation of player's role and player's self' (2000: 193).

Clearly there is a similarity here to Brecht's dialectical 'tussle and tension' between opposite registers of performance. Weimann's theory of *locus* and *platea* is, of course, rooted in the very specific cultural milieu of the Renaissance period; however, the central tension between the 'abstract and symbolic' register of the *locus* and the 'immediate and concrete' register of the *platea* is one which Colin Counsell, in his book *Signs of Performance*, has found 'useful for conceptualising modern theatre' (1996: 19), and one which he has particularly identified with what he calls the Concrete and Abstract registers of Brecht's theatre.[10] Since the Concrete/*platea* register shows itself 'in the act of telling', it foregrounds its own role (and by implication, the audience's) in the production of meaning.

There is another modern theatre practitioner with whom Counsell identifies a concern with the performance modes of *locus* and *platea*, however, and that is Peter Brook. For Counsell, Brook's use of the conflicting modes is opposite to Brecht: 'whereas Brecht proffered the *platea* and *locus* in contradiction ... Brook shows them as interdependent' (1996: 164). In Brook's use of spinning plates to represent the magical flower in *A Midsummer Night's Dream*, Counsell identifies the plate as the concrete *platea*, the fictional flower as the *locus*. But because of the skill involved in spinning the plate, Counsell argues, the *locus'* dependence upon successfully executed *platea* is openly displayed; 'The spectator must therefore acknowledge Concrete object and performer', says Counsell, 'and cooperate with him or her to build of the performance an other-place' – an interpretative relationship which Counsell describes as 'a semiotic collaboration' between actor and audience (1996: 164). We return, once again, to a fundamental divide between a popular theatre on the one hand which seeks to weave its disparate elements into a coherent unity, and on the other, one which seeks to emphasise its own fracturedness.

Shakespeare as popular theatre #2: A metatextual approach

What we might then term a 'Brookian' reading of the popular elements in Shakespeare is rooted in a liberal humanist understanding of the texts themselves; the discordant elements are brought into harmony to generate what Mitter describes as 'an authentic image of life's plurality' (1992: 5). Shakespeare uses conflicting registers, the argument goes, because life itself is made up of similarly varying modes of experience – in Brook's own words, 'The power of Shakespeare's plays is that they present man simultaneously in all his aspects' (1990: 98). Brook's productions are, by implication, bringing out what is already inherent in the plays in order to provide access to Shakespeare's timeless and unchanging truths about the human condition.[11]

This is a dehistoricisation of the texts quite at odds with contemporary critical theory, and, indeed, with Brecht. Brecht was categorical about the importance of treating old texts 'historically', which for him meant 'setting them in powerful contrast to our own time' (1965: 63–4). 'The emotions are in no sense universally human and timeless,' he argued elsewhere: 'the form they take at any time is historical, restricted and limited in specific ways' (1977: 145). Cultural materialist literary criticism shares his concerns. Alan Sinfield warns against the assumption that 'because human nature is always the same the plays can be presented as direct sources of wisdom' (2000: 190), and has criticised companies such as the RSC for

> productions which are intended to address political and historical matters, and which in some respects do that, but which at the same time make contradictory gestures towards a purportedly transcendent reality.
>
> (2000: 183)

Shakespeare's texts were not created in isolation, but rather as a result of the very specific historical, social, and political conditions within which Shakespeare and his contemporaries lived. Whether one reads the texts as having served the interests of 'the people' (as cultural materialists like Sinfield and Dollimore have done), as having reinforced the dominance of the status quo (as in Stephen Greenblatt's more pessimistic readings), or indeed as having staged a 'clash between different, often class-coded sentiments and ideals … out of which various voices of "the people" clamour to be heard' (Lanier 2002: 165), the fact remains that the plays are rooted in their own histories and that any attempt to make them appear unproblematically 'relevant' to modern audiences is guilty of denying the historical specificity not only of Shakespeare's age but also of our own. Steeped as we are in a theatrical tradition which looks always for the 'universal', particularly in Shakespeare, we are often resistant to the notion that the discourses we make Shakespeare speak in performance are never universal but always, in fact, constructed from our own social, political, and cultural concerns.

In this sense, popular theatre might provide us with a destabilising discourse, a means by which our habitually self-contained attitudes towards the texts might be disrupted. Since as we have seen, 'popular theatre' is often defined in opposition to whatever happens to consti-tute 'high' culture – and by any argument, Shakespeare must surely fall into the latter category – the fissure between 'Shakespeare' and 'popu-lar theatre' might become just as important as any correspondence between the two. In a debate with Richard Cuming in *Total Theatre Magazine*, Shakespearean scholar Rob Conkie defended a circus-style production he had directed of *The Comedy of Errors* at King Alfred's College Winchester as an attempt to 'destabilise the play' and to con-front 'a hegemonic representation of Shakespeare as elite and culturally edifying' (Conkie & Cuming 2003: 8). It is an impulse which can be seen in much popular Shakespeare. In the Shakespearean performances of the small-scale open air company Illyria, for example, anachronis-tic intrusions from pop culture – a rapping Stefano in *The Tempest*, or the inclusion of the *Teletubbies* theme in *Twelfth Night* – disrupt the patterns of spectatorship commonly associated with Shakespearean performance and force the audience into a playful reassessment of their relationship with the text. The provocatively irreverent work of companies such as Kneehigh and Vesturport will be explored later in the book.

Another means of interrogating the texts might be provided by popular theatre's interplay between *platea* and *locus*, between Concrete and Abstract. Here, though, unlike the approach which brings out the pluralism implicit within the texts, it is the split between Shakespeare's time and our own which is explored. Mitter's reading of Brecht describes the Concrete, *platea* register as a 'metatext', a new text *over-laid* onto the original; in Counsell's terms, 'The Concrete register takes an "attitude" towards the Abstract, offers meanings which conflict with it' (Counsell 1996: 105). Suggesting a cultural materialist attitude towards Shakespeare in performance along these lines, Simon Shepherd argues:

> We can play a scene so that the differences between performer and role become apparent. ... Cheek by Jowl's *Tempest* had a Black actress playing Miranda, who is usually played by a white woman. Automatically the person of the performer was separated from the role, while being fused with it in the narrative. The separation derived from the incomplete fit produced by racial difference, which in turn activated awareness of attitudes to race in the text.
>
> (Shepherd 2000: 228–9)

The capacity to effect such a split is, and always has been, much more evident in 'rough' theatre than in the 'legitimate'. In an interview conducted in 2006, director Annie Castledine explained to me that Complicite's exploitation in their production of *The Winter's Tale* of what she called 'the shining qualities of actors' skills' – the cast of nine played multiple roles, both comic and tragic – was in essence a Brechtian device. While the rare skill of being able to switch from play-ing the clown to playing the tragedian was, in her words, 'absolutely invigorating and inspiring' for audiences, it was designed, she said, to make it impossible for them to sustain any one response to the play (Castledine 2006).

Popular 'appropriations' of Shakespeare might go even further. A model, perhaps, is provided by Brecht's 'practice scenes for actors': one of these features a lovestruck Romeo forcing money out of one of his tenants; another reinvents Macbeth and Lady Macbeth as a Gatekeeper and his wife who cover up the breakage of an ornament belonging to their employer (Brecht 1967: 108–11). The Reduced Shakespeare Company, the Flying Karamazov Brothers and countless other popular theatre companies have produced satirical and deconstructive 'appro-priations' of Shakespeare's work; and in a theatrical culture where the primacy of the text is being challenged from all sides by physical and visual theatre on the one hand and postmodern performance on the other, the line which differentiates an 'appropriation' of a text from a 'faithful' interpretation is becoming increasingly blurred. This is explored further in Chapter 4.

In his introduction to *Shakespeare in Performance*, Robert Shaughnessy suggests that Brecht's 'dialectical theatre practice provides a practical model of political intervention in Shakespeare', and he throws down a challenge to Shakespearean theatre practitioners to bring cultural materialism's 'critical insights' to bear upon 'our practical negotiations with Shakespeare'. 'Whatever Shakespearean theatrical practice results,' he concludes, 'it is likely to be of a very different order to that currently on offer on the stages of the RSC or the Bankside Globe' (2000: 13). Douglas Lanier identifies a similar unfulfilled potential in what we have termed the 'collectivist' popular Shakespeare:

> As we have seen, this desire for an experience of communality typifies much of late-century popular Shakespeare performance. To be sure, that desire is often intertwined with nostalgia for some romanticised past and can be – indeed, often has been – placed in the service of Anglophilia, British nationalism, or various forms of elitism or exclusion. But it also speaks to popular discontent with the atomisation and alienation of modern life and contains within it inchoate forms of yearning for some social alternative, all of which might be harnessed to more progressive political ends. It is the challenge of popular Shakespearian performance to take up that potential, to prompt that imagined community within the Globe to engage the myriad inequities of the globe beyond its walls.
>
> (2002: 166–7)

This book will, I hope, show some of the ways in which contemporary Shakespearean and popular performances have begun to take up these challenges.

Personal Narrative 2
Stand-up Shakespeare

When I directed an outdoor production of *The Winter's Tale* for a UK summer tour in 2005, I encouraged many of the actors to make use of what I have described as *platea* modes of performance: direct address, audience interaction, colloquialisms, anachronisms, improvisation, and so forth. Nowhere did this seem more appropriate than in actor Dave Hughes' performance as Autolycus.

Passages of direct address are scripted for Autolycus throughout the text, employing what must, in the Elizabethan theatre, have been close to the vernacular language. In our version, therefore, Dave used these speeches as opportunities for short, entirely non-Shakespearean stand-up-comedy sets, making anachronistic allusions to such pop culture reference points as Harry Potter, *The Lord of the Rings* and *Murder, She Wrote*. Referring frequently to his battered Dover Thrift edition of the script, and to Charles and Mary Lamb's synopsis of the play, he provided an irreverent commentary on *The Winter's Tale* itself, at once both inside and outside the play.

When we approached the scene in which the disguised Autolycus torments the Clown and the Shepherd about the horrible fates which await them should they make themselves known to the King, we found that the character's knack for hurried and slightly surreal improvisation reminded us of the frantic ad-libbings of contemporary comedians such as Eddie Izzard or Ross Noble:

CLOWN.   Has the old man e'er a son, sir, do you hear, an't like you, sir?

AUTOLYCUS.   He has a son, who shall be flayed alive, then 'nointed over with honey, set on the head of a wasps' nest, then stand till he be three-quarters-and-a-dram dead, then

27

recovered again with aqua-vitae, or some other hot
infusion, then, raw as he is, and in the hottest day
prognostication proclaims, shall he be set against a brick-wall,
the sun looking with a southward eye upon him, where he is
to behold him with flies blown to death. But what talk we of
these traitorly rascals?

(4.4.782–92)

These influences in mind, I encouraged Dave to improvise a version
of this speech anew at every performance, incorporating his surroundings wherever he saw a humorous opportunity. Dave played the scene
disguised as 'Courtier Cap-à-Pie' in a monocle and a large waistcoat,
and spoke his lines with a manic energy and an exaggerated aristocratic
accent. Over the tour, his predicted tortures for the Clown ranged from
an *Indiana Jones*-inspired chase away from a giant spherical boulder
(in Canterbury) to an enforced spell of flyering on Edinburgh's Royal
Mile (at the Fringe Festival). The following is transcribed from a recording
of one of the final performances of the production in Sandwich, Kent,
where we performed on the Quayside, a few metres down the road from
the Sandwich Retirement Home:

CLOWN.   Has the old man e'er a son, sir, do you hear, an't like you, sir?

AUTOLYCUS.   Ah yes! He has a son, who shall be flayed alive, and then
he shall be sent to Sandwich! *(laughter)* But due to an admin-
istrative error, he shall be sent to the Sandwich Retirement
Home. *(laughter)* Yes … And there he shall spend a couple of
days consorting with the folk of the Sandwich Retirement
Home – *(faltering)* erm – with sandwiches – *old* sandwiches –
(laughter) – which are *three days* past their sell-by date! And
then he shall escape, because he does not like their conversation of lettuce and ham.*(laughter)* And what it was like when
they were freshly made.*(laughter)* So he shall run away! And he
shall go to Whitstable! *(laughter)* Where he shall *arrgh! (drops
monocle and frantically picks it up)* – where he shall join a travelling circus, and there he shall become a lion tamer! *(laughter)*
And using his knowledge of old sandwiches – which he has
gathered from the retirement home – *(laughter)* – he shall set up
a ring made of old sandwiches, and he shall command them to
run around! *(laughter)* But his show shall be a flop.

Because who wants to see old sandwiches running around? *(laughter)* And so he shall be fired, and put on the streets, and then – *(faltering again)* then a man with no arms will find him and say, *(miming no arms; increasingly staccato)* 'You must do this. ...' And he goes, 'What's this?' *(laughter)* And he goes, 'Go to the sea!' And he goes to the sea! But there are only flies at the sea! Because everybody knows there are lots of flies at sea. *(laughter)* For the sake of this story. And they shall blow him to death! *(panting with relief as he reaches an obvious cue line)* Blown to death with flies! But – *(laughter and applause)* – but what talk we of these traitorly rascals?

In an interview after the tour had finished, I asked Dave how he accounted for the generally very positive response that this part of the play received from audiences at all our tour venues. He explained it with reference to its unpredictability:

> When it started, I think I was so flustered and unsure of what I was saying, they [the audience] had to immediately pick up on the fact that maybe it wasn't scripted, and then as soon as they realised I was improvising, they really got hooked. Because it's interesting seeing people improvise on stage: there's that element that anything can happen.

Caitlin Storey, who shared the stage with him in this scene, made the additional point that the sequence momentarily realigned the other actors with the audience:

> It was interesting to see how far Dave could push this moment, and because we never knew how he was going to make it work each day; the cast, especially Martin [Gibbons] and myself, became members of the audience, joining together to laugh.

It is impossible to make an objective analysis of one's own work, but I would imagine there must also have been an element of transgression in the moment for audiences. A respect for the sanctity of Shakespeare's text is so automatic among theatregoers that a flagrant jettisoning of part of it in favour of haphazard improvisation must, I hope, carry something of a *frisson* of theatrical rule-breaking. The intention of such moments was not only to develop the elements of direct address and improvisation inherent in the script, but also to make a playful challenge to Shakespeare's cultural authority from outside it.

2
Text and Metatext: Shakespeare and Anachronism

(*A jug of mead is being passed around the circle. Hal makes as if to skip Azeem, the Moor.*)

ROBIN.   Has English hospitality changed so much in six years that a friend of mine is not welcome at this table?

FOLLOWER.   But he's a savage, sire!

ROBIN.   That he is. But no more than you or I. And don't call me 'sire'.

Robin Hood: Prince of Thieves (1991)

(*Robin has been affecting an American accent akin to Kevin Costner's in* Robin Hood: Prince of Thieves.)

MARIAN.   Robin, do stop talking in that silly voice. One of these days we're going to be famous, and our story will be told all over the world in moving picture galleries. It's going to look really stupid if we're all nicely spoken and you're gibbering on like a posey cowboy, isn't it?

Maid Marian and her Merry Men (1993)

Assimilative and disjunctive anachronisms

I open this chapter with passages from two recent retellings of the Robin Hood legend to illustrate two very different kinds of anachronism, both of which are common in popular narratives, but which have, in many respects, entirely opposite effects.

Both of these Robin Hoods are nominally set in the twelfth century, though clearly no attempt at historical accuracy is being made in

30

either case. The high-grossing, US-made *Robin Hood: Prince of Thieves* (dir. Kevin Reynolds) features some obvious, and probably deliberate, historical errors: its climax revolves around an anachronistic introduction of gunpowder, and Kevin Costner's use of his own American accent for the title role has become notorious (this is hardly a serious flaw, though, since Costner's dialogue in the film bears no less resemblance to the early English of the twelfth century than the received-pronunciation, cod-historical English spoken by even the most 'British' of screen Robin Hoods). But some of the film's anachronisms are more concealed. Pen Densham and John Watson's screenplay insinuates late twentieth-century ideologies into the legend, recasting Robin as an anti-racist, anti-sexist civil libertarian, centuries ahead of his time. According to this Robin, 'Nobility is not a birthright; it is defined by one's actions.' When Little John's wife Fanny picks up a sword to join the fight as the film approaches its climax, her husband protests that she should be minding the children. 'Tell her, Rob!' he says, turning to Costner's character for assent. But this Robin Hood rejects such old-fashioned values. 'Fanny,' he replies, like a true proto-feminist, 'you'll take position here'.

Tony Robinson's script for *Maid Marian and her Merry Men* (1989–94), on the other hand, turns this problematic relationship with the past into its greatest virtue. Broadcast on BBC1 between 1989 and 1994, and aimed primarily at an audience of children, Robinson's sitcom recasts Marian as the leader of the gang and Robin as a vain coward, playing with popular legend in a knowing and (dare I say it) almost postmodern fashion. References to the Costner film abound; playful acknowledgement of actor Howard Lew Lewis' minor role in the film is made when his character, gang-member Rabies, asks if he can be in a film of Robin's life, should one ever be made. Other anachronisms are shameless: a Rastafarian called Barrington is a key regular character; one of the local villagers invents the game of snooker; Much the Miller's son is reinvented as 'the Mini-Mart Manager's son'. When Robinson's conniving Sheriff of Nottingham speaks of his plans to hunt down 'Robin Hood and his robber band', his good-natured Brummie sidekick helpfully points out that 'rubber bands haven't been invented yet'.

The legend of Robin Hood is, I would like to suggest, analogous to the Shakespearean myth – a distinctive part of British national culture which lends itself equally to universalising romanticisation and to parodic appropriation and subversion. Both effects, paradoxically, can be brought about by anachronism. Of course, an element of anachronism is implicit in any project which attempts to retell the past; as Phyllis Rackin argues in *Stages of History*, 'the historian always constructs the past in retrospect, imposing the shapes of contemporary interests and desires on the relics of a former age' (1991: 94–5). This conflating of past and present is even more evident in popular appropriations of history, which by their nature tend to be aimed at non-specialist audiences and are often drawn from popular mythology rather than written history. Popular legends, whether in dramatic, narrative or ballad form, generally require their audiences to identify with the heroes of the story, and this will often lead to the kind of trans-historical blurring discussed above.

The way in which popular appropriations of history handle such anachronism can, however – as the Robin Hood examples illustrate – differ quite extremely. There are, essentially, two directly opposed forms of anachronism: those which attempt to hide, or at least underplay, the rifts between the present and a past historical period and those which seek to emphasise the same discrepancies. For the purposes of this study, I would like to call the former an 'assimilative' form of anachronism and the latter 'disjunctive'; the Robin Hood examples illustrate the two forms respectively.

Assimilative anachronisms are often conventional. It is so usual, for example, to hear contemporary English being spoken in historical settings (and in foreign ones – a kind of cultural 'anachronism') that we barely consider it anachronistic at all. The writer of the BBC's 2006 Robin Hood adaptation made the dubious claim in the *Radio Times* that '[m]ost of the time they [the characters] talk in believable medieval and we're trying to avoid faux-medieval' (Naughton 2006: 15); quite how 'believable medieval' differs from 'faux-medieval' in a twenty-first-century TV drama presumably not written in Middle English is anyone's guess. These anachronisms are 'back door' anachronisms: they creep into a historical narrative largely unnoticed and unremarked upon. The effect of this imposing of contemporary constructs on alien cultural settings, should it pass without comment, is homogenising and universalising; as Brecht put it,

> Whenever the works of art handed down to us allow us to share the emotions of other men, of men of a bygone period, different social classes, etc., we have to conclude that we are partaking in interests which really were universally human.
>
> (1977: 146)

But as Brecht argues earlier in the same essay,

> The emotions always have a quite definite class basis; the form they take at any time is historical, restricted and limited in

specific ways. The emotions are in no sense universally human and timeless.

(1977: 145)

The illusion of trans-historical empathy discourages us from seeing the ways in which past periods were shaped by ideologies unfamiliar to us today. By implication, we are encouraged to see our own values as 'universal' rather than constructs which are very socially and culturally specific; to quote Brecht once again, 'to say that something is natural ... is to / Regard it as unchangeable' (1990: 37).

Disjunctive anachronisms, on the other hand, deliberately break with convention. Where assimilative anachronism homogenises past and present, this kind of anachronism disrupts, deconstructs, and destabilises. Where the former buries the imaginative impositions we make upon the past, the latter flags them up; our attitudes towards the past are shown frankly (and often humorously) to be the products of the culture in which we live. In John Madden's 1998 film *Shakespeare In Love*, for example, our first sight of the title character is a culturally familiar one: Joseph Fiennes' Shakespeare sits in an attic room, scribbling away with a quill, throwing discarded pieces of paper over his shoulder – presumably, it would appear, channelling his Muse. But our romanticised image of the character is immediately disrupted when the camera shows us that Shakespeare is not composing verse, but practising his signature. The camera then pans across to show us the amusingly anachronistic inscription on Shakespeare's mug: 'A present from Stratford-upon-Avon'.

The film's playfully anachronistic attitude pervades its opening sequence. Shakespeare visits 'Doctor Moth' (Antony Sher), whose outside wall bears the placards 'Priest of Psyche' and 'Interpreter of Dreams'. Inside, Shakespeare lies down on what looks very much like a modern therapist's couch, and confides his psychological torment. Passages like the following, which relocate clichés of contemporary life (in this case, the talkative taxi driver) into the Elizabethan setting, implicitly acknowledge that the film can never be anything but a very dubious reconstruction of the past, fashioned in the image of the present:

WILL.   Follow that boat!

BOATMAN.   Right you are, governor!
     WILL sits in the stern of the boat and the BOATMAN sits facing him, rowing lustily.

BOATMAN.   I know your face. Are you an actor?

WILL.   (*oh God, here we go again*) Yes.
BOATMAN.   Yes, I've seen you in something. That one about a king.
WILL.   Really?
BOATMAN.   I had that Christopher Marlowe in my boat once.

(Norman & Stoppard 1998: 36–7)

One might be forgiven for suggesting that this sounds like the province of postmodernism, but in fact disjunctive anachronisms of this kind have been employed in popular narratives for centuries. Weimann argues that such dramatic anachronism is 'characteristic of the popular tradition' (1987: 80): folk dramas and mystery plays would, as we saw in the opening chapter, make use of an interplay between *locus* and *platea* and their corresponding modes of performance to set the present in stark contrast with the past.[1] The *Second Shepherds' Play* of the Towneley cycle names the Shepherds of the nativity story Coll, Gyb, and Daw, and adds the characters of Mak the sheep-stealer and his wife Gyll. These are clearly inhabitants of fifteenth-century England rather than biblical Bethlehem: they express medieval attitudes and swear anachronistically by Christ, by Saints James and Stephen, and even by 'Thomas of Kent'. But it is not, argues Weimann, 'the absence of a historical perspective' that lies behind these kinds of anachronism:

> rather, a positive principle is at work, according to which the actor may achieve both a specific degree of approach to, and dissociation from, the audience's world in order to move freely between the poles of sub-jectivity and objectivity, self-expression and representation. ... Thus, anachronism points to the contradiction between biblical myth and medieval reality as well as to the connection between subject-matter and interpretation, pathos and burlesque, seriousness and comedy.
>
> (1987: 83)

Once again, a theatrical model similar to Brecht's is suggested here. Disjunctive anachronisms serve a discursive function – they contribute to a multiplicity of perspectives. Colin Counsell describes Brecht's Epic theatre in contrast to the 'g-werk', the kind of theatre which 'presents itself as authorless and non-constructed, its meanings not provisional but a given truth' (1996: 105):

> Epic theatre in contrast advertises its own fictionality, for overt arti-fice reveals an author's hand at work behind the text's constructions.

Epic theatre always shows itself in the act of telling, and its arguments are presented as manufactured. Whereas the g-werk is separated from the audience's social space, signs of artifice link the playworld to that space, show its meanings as produced within it.

(1996: 105)

As we saw in the opening chapter, Counsell identifies this as a form of interplay between *locus* and *platea*.

Disjunctive anachronism continues to thrive in popular culture today.[2] We might see a 'medieval laundrette' scene in a pantomime, or watch *The Flintstones* visit a stone-age bowling alley. Recent popular films such as *Moulin Rouge!* (dir. Baz Luhrmann , 2001) and *A Knight's Tale* (dir. Brian Helgeland, 2001) revel in their anachronism: the former allows *fin de siècle* Parisians to break out into late-twentieth-century pop songs, while the latter features scenes in which the crowd at a medieval jousting match sing and hand-clap along to 'We Will Rock You', and a female blacksmith adorns her armour with Nike 'ticks'. Such anachronisms function as satirical commentaries on the foibles of contemporary life at the same time as playfully flagging up the fundamental impossibility of creating an accurate dramatic representation of the past.

Films such as Julie Taymor's *Titus* (1999) take this disjunctive anachronism a step further. In the first scene, a child playing with toy soldiers and aeroplanes in a recognisably late-twentieth-century kitchen is dragged out into a Roman amphitheatre by a man in motorcycle goggles; shortly afterwards, a nightmarish procession of Roman-style soldiers enters the amphitheatre, flanked by Roman chariots and Second-World-War motorcycles. Alan Cumming's Saturninus, with his theatrically impassioned speech to the citizens of Rome, his military outfits, and his starkly sideswept hairstyle, echoes images of Adolf Hitler; Jonathan Rhys Meyers and Matthew Rhys's camp and vaguely futuristic Chiron and Demetrius, meanwhile, recall Alex and his 'droogs' from Kubrick's film adaptation of *A Clockwork Orange*. Taymor's film frustrates any attempt to locate the story historically, constantly unsettling and dislocating its audience, drawing historical parallels one moment only to tear them down the next. Both history and narrative are shown to be forms of discourse.

Of course, as with most of the polar oppositions described in this book, an element of both assimilative and disjunctive anachronism can be detected in most popular appropriations of history. Weimann, it should be noted, describes 'a specific degree of approach to, *and* dissociation from, the audience's world' (1987: 83; my emphasis). As we saw in the first chapter, inconsistent and contradictory attitudes can exist without synthesis in the same piece of theatre, and as I have already suggested, this disjunction can lend a discursive aspect to the play's effect upon its audience. Certainly Shakespeare's own use of anachronism, as we shall see in a moment, spanned both categories.

But this chapter is not concerned only with Shakespeare's use of anachronism. In the first chapter, I described two opposing approaches towards appropriating Shakespeare's work for 'popular' performance today: the 'textual', which suggests a relatively 'faithful' rendering of Shakespeare's text, and the 'metatextual', in which a new, non-Shakespearean text is 'overlaid onto the original'. As we have already seen, a certain measure of anachronism is inherent in the very idea of a 'metatext'; this can, however, be either assimilative or disjunctive. To sum up, then, it may be possible to identify four distinct approaches to anachronism in contemporary Shakespearean performance:

1. **Textual/Assimilative:** Most contemporary mainstream productions tend to overlook Shakespeare's own anachronisms. Anachronistic clown characters, for example, are thus integrated seamlessly into a historical setting, and the *locus-platea* interplay implied in the texts is understated.
2. **Textual/Disjunctive:** Some productions will emphasise and elaborate upon the disjunctive anachronisms inherent in the text, for the primary purpose of fulfilling the textual functions of these anachronisms – what W. B. Worthen, borrowing the term from Joseph Roach, calls 'surrogation'. 'An act of memory and an act of creation,' he explains, 'performance recalls and transforms the past in the form of the present' (Worthen 2003: 64).
3. **Metatextual/Assimilative:** Many productions will add their own assimilative anachronisms (a non-Shakespearean setting, for example), as a means of assimilating the text to contemporary values. This is often characterised as highlighting Shakespeare's 'relevance', and is common practice at the RSC.
4. **Metatextual/Disjunctive:** Occasionally, a production will add its own disjunctive anachronisms as a means of problematising or deconstructing its own relationship with the Shakespearean text.

As I have implied, these categories are, in practice, never this clear-cut; a purely 'textual' approach is, in reality, impossible to sustain (the 'metatext' will always contaminate the production of meaning, even if only accidentally). In emphasising Shakespeare's own anachronisms, for example (what I have called a textual/disjunctive approach), it will be almost impossible by the same token to avoid some sense of disjunction from the text itself. Examples of this will be explored later on; but as with the *locus-platea* interplay explored in the first chapter, it is often the contrasts between a combination of these approaches within a single piece that produce its overall effect.

Textual anachronisms

In his essay on 'Shakespearean anachronisms', Jonas Barish cites the following examples of anachronism in Shakespeare's work:

> Hector quoting Aristotle – several centuries before Aristotle was born; the future Richard III, while Duke of Gloucester, measuring his own ruthlessness against that of the murderous Machiavel – at a time when Machiavelli was still in his infancy; Hamlet attending an as-yet-unfounded Wittenberg University; Cleopatra playing billiards.
>
> (1996: 29)

'These are the kinds of error', Barish concludes, 'from which our most universally revered culture hero seems not to have been exempt' (1996: 29). But Barish's list is noticeably confined to anachronisms which are (with the possible exception of the last) not immediately obvious as such to the average playgoer. As Barish himself points out,

> if we, with our historical noses to the ground, do not scent historical falsity in these cases, surely no Elizabethan theatregoer, far less schooled than we in detecting historical discrepancies, would have noticed anything amiss, or would have cared two pins if he had.
>
> (1996: 31)

These examples (again, with the exception of the example from *Antony and Cleopatra*) illustrate a very minor sort of assimilative anachronism, in which characters assume a knowledge or experience of history which would have been available to the Elizabethan audience, but which is strictly inaccurate for their own historical settings. Since they draw no anachronistic parallels with the Elizabethan present, however, their assimilative effects are likely to be limited, and since they are not noticeable enough to a non-specialist audience to cause any kind of historical dislocation, they cannot be said to be disjunctive either. Rather, they might be described simply as historical errors, and relatively unimportant ones at that.

Comparing Shakespeare's *Antony and Cleopatra* with Chapman's *The Blind Beggar of Alexandria* (1598) – a play which features pistols, tobacco, Christian references, and modern Spanish in an Ancient Egyptian setting – Barish points out that while Chapman's play 'simply rides roughshod over all considerations of chronological plausibility, and glories in doing so', Shakespeare tends to offer 'very few examples of such wanton flouting of temporal plausibility' (1996: 30–1). I wonder, however, whether Barish is not overlooking some very important and deliberate anachronisms here. Naturally it is impossible to identify exactly to what extent the average Elizabethan playgoer would have identified the various anachronisms in Shakespeare's plays; probably, in many cases, there was no 'average' response at all, with some playgoers spotting anachronisms where others remained oblivious. But while Barish suggests that Shakespeare 'was plainly aiming at a persuasive recreation of older cultures' (1996: 31), I would argue that Shakespeare's texts contain many anachronisms which suggest the very opposite: a disruption of historical narratives with intrusions from the Elizabethan present.

This is not to claim that there are no significant assimilative anachronisms in Shakespeare's plays. The aforementioned billiards in *Antony and Cleopatra* (2.5.3), the clock which strikes in *Julius Caesar* (2.1.192 and 2.2.114), and the doublet, hat and hose which Pisanio gives to Imogen in *Cymbeline* (3.4.170), are all examples of anachronisms which imply a false continuity between the ancient past and the Elizabethan present. More significantly, most of Shakespeare's historically-set plays display anachronistically Elizabethan world views; *Macbeth*, for example, ascribes Elizabethan concepts of kingship and nationhood to eleventh-century Scotland, where power systems and the ideological constructs which supported them were in reality very different. But these assimilative anachronisms work alongside and in contrast with such decisively disjunctive anachronisms as the Elizabethan clowns with unambiguously English names and contemporary occupations – Nick Bottom the weaver, Francis Flute the bellows-mender, Tom Snout the tinker, and so on – who inexplicably inhabit ancient Athens. In a similar vein, culturally if not temporally anachronistic characters such as Anthony Dull, Susan Grindstone, Kate Keepdown, Hugh Oatcake, and George Seacole populate settings as diverse as Navarre, Verona, Vienna, and Sicily.

One might argue that the Elizabethan audience's familiarity with the anachronistic clown character would have rendered the device conventional (and therefore assimilative) rather than disjunctive, but the texts are often at pains to point out their own anachronisms. The historical impossibility of the Fool's cryptic exit line in the storm scene of the Folio *King Lear* must have been obvious to even the least educated audience member, since it draws deliberate attention to itself: 'This prophecy Merlin shall make,' says the Fool, 'for I live before his time' (3.2.95). Similarly, when Cassius in *Julius Caesar* speculates over Caesar's body as to the future-historical consequences of his actions, he phrases it with self-advertising anachronism:

> How many ages hence
> Shall this our lofty scene be acted over,
> In states unborn and accents yet unknown!
> (3.1.112–14)

Cleopatra expresses her fears of an as-yet-unheard-of dramatic tradition presenting her story in a similar way:

> The quick comedians
> Extemporally will stage us, and present
> Our Alexandrian revels. Antony
> Shall be brought drunken forth, and I shall see
> Some squeaking Cleopatra boy my greatness
> I'th' posture of a whore.
> (5.2.212–7)

Since the Elizabethans would have seen a drunken representation of Antony in 2.7, and indeed witnessed a 'squeaking boy' speak these very lines, the whole passage is steeped in a dramatic irony which is clearly disjunctive.

A certain degree of overt anachronism was evidently central to the Elizabethan theatre. Henry Peacham's illustration of the opening scene of *Titus Andronicus* in the Longleat manuscript (c. 1595) shows a mixture of Roman and Elizabethan costume: the figure presumably representing Titus wears Roman apparel, while Tamora and the guards wear more anachronistically Elizabethan attire. While it is inadvisable to base too much supposition upon the evidence of a single drawing, one might infer from this that those characters assuming greater historical significance were more likely to adopt the associated historical dress, while in other characters, historical accuracy was of less importance. As Rackin suggests, 'women and commoners have no history because both are excluded from the aristocratic masculine world of written historical representation' (1991: 103). But the Longleat illustration, argues Rackin, is merely one example of a dichotomy which permeates Shakespeare's theatre:

> Shakespeare locates his highborn men in a variety of historical worlds, but his commoners belong to the ephemeral present moment of theatrical performance, the modern, and socially degraded, world of the Renaissance public theatre.
>
> (1991: 103)

Rackin's observation echoes Weimann's analysis of Shakespeare's anachronisms. On the Renaissance stage, Weimann argues, 'it is invariably the figures close to the audience, especially the clowns and fools, who perpetuate the tradition of anachronism' (1987: 80). These characters, descended (to varying extents) from the Vice character of the morality play, would often provide an anachronistic intrusion into the play's historical setting for the purpose of puncturing its status as 'official history', providing a satirical counter-perspective. 'Invoking a degenerate present and a disreputable scene of theatrical performance to degrade the idealised past,' argues Rackin, 'Shakespeare's anachronisms usually function as tokens of debasement' (1991: 104). Thus the clown who undermines the grandeur of Saturninus' court with his pigeons in *Titus Andronicus* strikes an anachronistic note, both with his ignorance of Jupiter (4.3.84–5) and with his reference to 'God and Saint Stephen' (4.4.42); while the latter is not strictly anachronistic according to the play's unspecific late–Roman Empire setting, it certainly belongs more pronouncedly to the present of the Elizabethan audience. *Antony and Cleopatra*'s distinctly Elizabethan clown, meanwhile, drags the play's climax suddenly into a comically mundane prose discussion of 'the worm' (5.2.240–74), while the intrusion of the cynical Autolycus in Act Four of *The Winter's Tale*, selling folk ballads, cheating peasants, and satirising royalty, is markedly at odds with the play's classical setting. The speech patterns of such characters, written in prose and rooted in the vernacular of the Elizabethan present, were themselves anachronistic, as Weimann explains

> The clowns could, at least on one level, understand the rhetorical, stylised speech of central characters, but on another they could intentionally misunderstand it, invert it, garble it, or incorporate it

into wordplay that revealed either a more deeply knowing or a naive complementary view.

(1987: 244)

But the legacy of the morality Vice character was not limited to the prose clowns. Edmund, Iago, Aaron, and Richard III all address their audiences directly, implying the playgoers' collusion as they confide their wicked plans. As such, since Richard makes his reference to the 'murderous Machiavel' (*3 Henry VI*, 3.2.193) while he is talking directly to an Elizabethan audience (themselves an anachronistic intrusion into the fifteenth-century setting), it can hardly be taken seriously as a grave historical error: an element of overt anachronism is so inherent in the character's dramatic pedigree here that to worry about this minor example of it would be much like quibbling over an anachronistic use of, say, Roman weaponry in *Monty Python's Life of Brian*.

Shakespeare's layering of conflicting historical registers, like the interplay of *locus* and *platea* discussed in the first chapter, serves to emphasise history as a form of discourse. Discussing *Cymbeline*'s anachronistic (one might even say ahistorical) leaps from one period of history to another – Renaissance Italy, for example, seems to co-exist with Augustus Caesar's Rome – Huw Griffiths suggests that 'anachronism here functions as a Brechtian moment in which the stage is revealed as a construction of the past rather than an accurate record of it' (2004: 356). Rackin's chapter on anachronism in *Stages of History* explores such layering more fully; identifying the Boar's Head scenes in *Henry IV* as anachronistic (1991: 103), she explains that their 'references to the present scene of theatrical performance' serve to subvert what she calls the 'historiographic project of ideological mystification' implicit in the scenes where medieval kings and aristocrats 'quarrel in blank verse and play out the scripts written by Tudor historiography' (1991: 98; 138). Official histories are thus opened up to satirical deconstruction. Rackin argues that in Pistol, 'a character whose very name is an anachronism' (1991: 139),

[Shakespeare] uses a stock character from the sixteenth-century stage to mark the anachronistic distance that separates the historically specific site of his own theatrical representation from the lost past it can never truly present.

(Rackin 1991: 143)

It is the self-consciously theatrical Pistol who brings the news of Hal's coronation to the stage at the end of *2 Henry IV*; it is also Pistol who is the sole survivor of all the Eastcheap characters at the end of *Henry V*, when, unreformed and embittered, he resolves, 'To England will I steal, and there I'll steal' (5.2.83). The similarity to the Vice character is marked here, and it is significant that Pistol – provider of a cynical and anachronistic perspective on the king's actions throughout *Henry V* – is allowed to survive while all his comrades (Falstaff, Bardolph, Nim, Mistress Quickly) perish.

Perhaps the reason that these very conspicuous anachronisms are often overlooked by critics lies in the history of theatre itself. Already in Shakespeare's time, the use of contradictory dramatic registers was being denounced by such critics as Dr Joseph Hall, who deplored the stage's 'goodly *hoch-poch*, when vile *Russettings*, / Are match'd with monarchs, & with mighty kings',[3] and Sir Philip Sidney, whose condemnation of plays that were 'neither right Tragedies, nor right Comedies, mingling Kinges and Clownes' is famous.[4] By the late eighteenth century, the emphasis on a neoclassical unity of tone was so pronounced that the very idea of audience members being a visible part of the 'background' to a scene like the one at the Capulet vault in *Romeo and Juliet* caused the actor Tate Wilkinson to observe, 'I do not think at present any allowance but peals of laughter could attend such a truly ridiculous spectacle'.[5] Similar attitudes towards anachronism accompanied this development in theatrical taste; Alexander Pope, according to Barish, was so 'horrified' at the reference to hats in *Julius Caesar* (2.1.73) that in his 1725 edition of the play, 'rather than attempting any emendation, he left a blank in the text, as though deleting a foul expletive' (Barish 1996: 33). In 1765, Samuel Johnson's Preface to his edition of *The Plays of William Shakespeare* constructed Shakespeare's characters as unmodified 'by the accidents of transient fashions or temporary opinions'; they were, he claimed, 'the genuine progeny of common humanity, such as the world will always supply, and observation will always find' (1965: 12). Assimilative anachronism was overlooked, then, at least by Johnson: disjunctive anachronism, however, was a different matter. In the same edition, Johnson complained of *Cymbeline* that to comment on the play's 'confusion of the names and manners of different times' would be to 'waste criticism upon unresisting imbecility, upon faults too evident for detection, and too gross for aggravation' (1965: 183). For Johnson, Rackin surmises,

> the anachronisms can only be faults, either faults to be blamed as the embarrassing evidence of Shakespeare's lack of education or faults to be excused as the products of an unenlightened age or the by-products

of a genius too preoccupied with the essence of universal truth to trouble itself with the accidents of transient fashions or temporary opinions.

(1991: 88)

These positions, she concludes, 'pretty much exhaust the range of commentary on Shakespearean anachronism from Johnson's time to our own' (1991: 88). Indeed, the nineteenth-century Naturalist movement in the theatre has left us with an overinflated sense of the importance of historical accuracy in drama which has only recently started, once again, to be questioned.

Metatextual anachronisms #1: Assimilative approaches

When Sir Barry Jackson presented his 1925 production of *Hamlet* for the Kingsway Theatre in modern dress, the *Times* described the device as an 'experiment' and an 'eccentricity', remarking that

[w]hen the curtain rises on the Court in evening dress and Orders there is, and must be until we are accustomed to the convention, a sensation of shock; for a moment the mind is turned away from what is important to the non-essentials – to the fact, for instance, that the music of the dance is syncopated or that the courtiers drink cocktails and light cigarettes. And question is – it is the question by which the whole experiment must stand or fall – how long will this diversion last? ... The answer is that the diversion is short. It will recur now and then, later in the play. ... The initial strangeness vanishes, leaving a clear gain in freshness and life and vigour, in almost everything except that visible beauty which the eye demands and cannot here discover.

(*Times*, 26 August 1925)

Modern dress is so usual in Shakespearean production today that one cannot read the *Times* critic's warning that 'the experiment, though it has in this instance proved its substantial value, is not one that others should lightly dare repeat' without some amusement. Clearly in 1925, modern dress was a form of disjunctive anachronism, undertaken as a dramatic innovation. Today, however, contemporary or otherwise anachronistic costume is generally accepted as a key tool in making Shakespeare 'relevant', and has thus assumed a much more 'assimilative' role.

The widespread desire to emphasise the continuing relevance of Shakespeare's plays has led mainstream Shakespearean productions today to adopt a variety of approaches towards the theatrical 'metatext' (that is, those aspects of the production which are imposed upon the play from outside it) which might be described as assimilatively anachronistic. First, as suggested above, is the kind of historical relocation which has become typical of productions by the RSC, the National or the Old Vic. In these productions, the Shakespearean play is lifted without any substantial textual alteration into an alien historical setting, and the new setting reveals an aspect of the play's contemporary significance: thus David Thacker's RSC production of *The Merchant of Venice* in 1993 adopted a 1980s city office setting in order to emphasise the play's insights into the more ruthless aspects of capitalism, while Trevor Nunn's production of the same play for the National Theatre in 1999 assumed a European jazz-age setting reminiscent of Kander and Ebb's *Cabaret*, confronting the difficulty of staging the play's anti-Semitism in a post-holocaust world. In a similar way, *Much Ado About Nothing* became the tale of an unequal and decadent society on the brink of collapse in Marianne Elliott's 2006 RSC production, set in 1950s Cuba; Trevor Nunn's *Richard II* at the Old Vic in 2005 presented a cold war being fought through the manipulation of the twenty-first century media; Gregory Doran's *Othello* became a story about colonial (and by implication, postcolonial) race relations in his 1950s Cyprus setting.

Of course, Shakespeare himself wrote about none of these things. Any insights he had into political manipulation, race relations, or the workings of capitalism, must necessarily have been limited to the ways in which those issues (or their historical equivalents) impacted upon Elizabethan society, and while those insights may be pertinent and persistent today, they must be historically, socially, and politically located. In appearing to fit Shakespeare's plays seamlessly into wildly different historical settings, the anachronisms of the productions discussed above make the implication, as we saw in the first chapter, that Shakespeare's work provides access to unchanging truths about the human condition.

Shakespeare's continuing trans-historical relevance is insinuated in other ways too. The imposing of contemporary speech patterns upon the verse, mimicking the cadences of naturalistic dialogue, is a common trick (though it can often have the counterproductive effect of making the poetry more difficult to understand).[6] And even when extra-dramatic, non-Shakespearean intrusions threaten to disrupt the narrative, the Shakespearean text often displays an odd ability to accommodate them: the plays' frequent references to the wind, rain, and sun are often met with laughs or comic groans in performances at Shakespeare's Globe, for example, when they appear to describe either the present weather conditions, or their exact opposite. More striking are examples such as the anecdote related by David Warner about his celebrated 1965 *Hamlet*: as he delivered the 'O, what a rogue and peasant slave am I' soliloquy (2.2.551–607), upon reaching the line, 'Am I a coward?' he was met with the answer, 'Yes!' from a member of the audience. Scanning the audience for his critic, Warner responded with the next lines, finding them surprisingly apt:

Who calls me villain, breaks my pate across,
Plucks off my beard and blows it in my face,
Tweaks me by th' nose, gives me the lie i'th'throat
As deep as to the lungs? Who does me this?

Now Warner/Hamlet's question was answered a second time – the man shouted out his name. Once again, Warner found Hamlet's response remarkably appropriate. 'Ha?' the character responded; ''Swounds, I should take it ...'[7] Naturally examples such as these appear to testify to the plays' abilities to transcend historical specificity, but all they really indicate is a persistent potential for reinvention and reinterpretation.

Many directors justify their metatextual additions to the Shakespearean texts with reference to 'what Shakespeare intended'; had he been alive today, the argument goes, this is what he might have done. Trevor Nunn explored the idea in an article for the *Guardian* under the subtitle: 'Many theatregoers call for "Shakespeare as Shakespeare intended" – but what is that exactly?' (2005: 14). Though Nunn's article does acknowledge that 'we cannot be sure of Shakespeare's intentions', he does imply that Shakespearean productions noted for 'their daring, their unexpectedness and their determination to provide fresh insight, new stimuli and increased relevance' are in some way fulfilling Shakespeare's wishes. Nunn defends directorial reinterpretation and anachronism (and by implication, the inherent anachronism of his own concurrent production of *Richard II*, in which the dialogue of Shakespeare's medieval kings and aristocrats was spoken by media celebrities at photoshoots, and sharply dressed politicians via live video feeds) on the grounds of a Shakespearean precedent:[8]

We know that he packed his work with contemporary references and satirical portraits of the rich and famous, and that he was never bothered by anachronisms. Cleopatra playing 'billiards'; an ancient Briton mocked as a 'base football player'; Pistol, living in early 15th-century London, characterised as a regular at the Playhouse – these were all part of Shakespeare's preference to keep his audience colloquially involved and in a state of spontaneous recognition rather than satisfied scholarship.

(2005: 14)

Nunn's argument is a persuasive one, but it is interesting that he needs to make it at all. Since the publication of Barthes' influential essay 'Death of the Author' in 1967, it has become increasingly unnecessary to justify an interpretation of a text with reference to the author's wishes: 'To give a text an Author,' Barthes argued, 'is to impose a limit on that text, to furnish it with a final signified, to close the writing' (1977: 147). In any case, the dramatic text does not contain some 'pure' version in and of itself, which is somehow truer than any theatrical interpretation; it has no meaningful existence until it is brought to life either in the theatre or in the mind of the reader. And in the theatre, as David Wiles has argued, an insistence upon a strictly 'textual' reading is impossible:

in addition to the literal heteroglossia provided by the multiple voices of the actors, and the more subtle heteroglossia provided by the different linguistic registers in the dialogue, we also have the language of gesture which may speak against the words.

(1998: 67)

As Wiles' argument implies, a certain level of 'heteroglossia' is inevitable in practical performance; a theatrical 'metatext' is unavoidable. The almost god-like status ascribed to Shakespeare by what might be called the 'Shakespeare Industry' seems, however, to downplay the role of this necessarily multivocal metatext and insist upon the author's dominance.

Alan Sinfield has argued that in fact this two-pronged ambition to make Shakespeare 'relevant' and at the same time be seen to be fulfilling his intentions is made problematic by the inherent contradiction between the political specificity demanded by the first condition and the ahistorical transcendence implied by the second. The logic involved, he argues, is circular:

The central idea of Shakespeare, so far from affording some control over what it is that the plays might represent, is actually used to justify at least sufficient interpretive scope to secure relevance. Then, conversely, the relevance of any particular production is guaranteed by the fact that it is Shakespeare, who is always relevant. The circle seems unbreakable.

(2000: 187)

Naturally, 'Shakespeare's intentions' can never be objectively proven; they will always be nothing more than a set of constructions. If a production strays too far from what Sinfield describes as the 'conventional under-standing of the plays', then it can be seen as seeking relevance at the expense of textual fidelity: 'if you push the Shakespeare-plus-relevance combination too hard,' he suggests, 'it begins to turn into a contradic-tion' (2000: 188). Sinfield gives Jonathan Miller's 1970 *Merchant of Venice* as an example of this; Miller was accused in the press of privileging his own ideas over Shakespeare's. Bearing in mind the manifest anti-Semitism presented in the play, one might wonder why this was consid-ered a problem.

The Merchant of Venice provides perhaps the most striking examples of Sinfield's observation. In a 1994 debate with the playwright Arnold Wesker, David Thacker defended his production of the play from Wesker's accusa-tion that he had 'invented a fantasy about what you would have liked Shakespeare to have written'.[9] Thacker tried to hold down the mutually contradictory positions that he was at once deconstructing Shakespeare's play and attempting to fulfil Shakespeare's wishes; he claimed, 'I go on the imaginary premise that if Shakespeare came back to life, what would he do?', but suggested at the same time that he was deliberately distancing himself and his production from Shakespeare's text:

WESKER.   This production can be accused of being dishonest. I
heard of a production in Hungary during the war where
they really played it for what it was and it worked as a Nazi
tract. That's more honest. What you've done is cut the text
around to make it fit a notion which you feel very strongly
about.

THACKER.   You're absolutely right. It's dishonest in that it is
subverting the text.

WESKER.   Let's be more generous and say it's not dishonest, it's not
consistent.

THACKER.   It's clear these people [the characters] are bigoted, they
hate gays, Jewish people, black people. The alteration of the

trial scene is so that the final image you get is of Shylock standing facing another day. Shakespeare's words, just in a different order ...

WESKER.   I think that my wish to write a new play [Wesker's *The Merchant*, later called *Shylock*, 1977, published as *The Merchant*, 1978] is a more honest approach to the problems than re-jigging and imposing on the play.

THACKER.   We've taken the same choice as you took in your play.

WESKER.   There's a complete difference. I have not adapted Shakespeare's play. I've used the same three stories to write a completely different play.

(Armitstead 1994: 4)

If Thacker had accepted Wesker's accusation of 'imposing' on the play, his argument might have been stronger: after all, it is impossible to produce Shakespeare's plays today *without* imposing a contemporary perspective. As it is, though, Wesker's allegation of inconsistency sticks: Thacker's reluctance to admit his production's tentative departure from Shakespearean 'authority' ultimately undermines his claim that the production was 'subverting the text'.

Metatextual anachronisms #2: Disjunctive approaches

Wesker's central criticism of Thacker's production was that its 're-jigging and imposing on the play' was somehow dishonest. This is arguable: a production relocating the play into a 1980s setting can hardly be said to be hiding its status as an 'interpretation', but as we have seen, productions such as Thacker's frequently 'make contradictory gestures towards a purportedly transcendent reality' (Sinfield 2000: 183). A more 'honest' approach, then, might be to advertise the discontinuity between text and interpretation quite unashamedly. This necessarily entails a certain deliberate distancing from any notion of 'Shakespeare as Shakespeare intended', and a disruption of any kind of humanist reading of Shakespeare as provider of universal truths.

As we might expect, this disjunctive approach towards Shakespeare is likely to be found much more pronouncedly in the more 'popular' productions of the texts; indeed, as we saw in the first chapter, a disjunctive approach is likely to contribute towards *defining* a production as 'popular', since it implies a split from the 'legitimate' Shakespearean tradition.

Such anachronism often takes the form of textual interpolation: lines are changed, topical references or contemporary phrases added, and self-reflexive jokes made around the text of the play. Such strategies are, on the whole, avoided in mainstream Shakespeare, where only non-verbal interpolations ('aahs' and 'oohs', for example) are generally seen as compatible with 'textual fidelity'. The reasoning, such as it is, appears to be that non-textual interpolation will pass unrecognised as such just as long as no words are spoken; thus, for example, characters' unscripted objections are often somewhat awkwardly signified by an 'uh-uh-uh' and a wag of the finger. The device can be taken to extremes: in Nancy Meckler's RSC *Comedy of Errors*, Christopher Colquhoun's Antipholus of Ephesus turned his physical illustration accompanying the line 'My wife is shrewish when I keep not hours' (3.1.2) into a fully blown mime sequence, culminating in an impression of *Psycho*'s Norman Bates raising his dagger, accompanied by a vocal recreation of Bernard Harmann's famous screeching violins. That it was a non-textual interpolation is clear; but since it featured no verbal departure from the script, it seemed to obey the RSC's unspoken rule.

Verbal interpolations, while rare, are not however completely unheard-of in mainstream Shakespeare. In Gregory Doran's 1999 *Macbeth*, for example, Stephen Noonan's confrontational Porter padded out the comic sequence with topical knock-knock jokes and references to BSE and hospital waiting lists; his 'equivocator', meanwhile, became a Tony Blair who was 'totally committed to the notion of equivocation'. This might be defined as an example of 'textual/disjunctive' anachronism; paradoxically, Noonan was fulfilling some of the textual functions of the Porter's role (providing a sudden change in perspective, a contrast in mood from the treachery and tension of Macbeth's Dunsinane, and suggesting contemporary parallels with Macbeth's duplicity) by departing from the text itself. Complicite's 2004 *Measure for Measure* at the National Theatre, meanwhile, featured such interpolations as Escalus' patronising 'Thanks for your input; take a seat,' to Elbow, and Pompey's 'Mind the cat' (accompanied by a vocal impression of a cat being trodden on) as he led another character along a mimed corridor. Most noticeably disjunctive was an exchange between Richard Katz's Pompey and Tamzin Griffin's Mistress Overdone:

MISTRESS OVERDONE.   What, is there a maid with child by him?

POMPEY.   No, but there's a woman with maid by him. *(to audience)*
It's Elizabethan, don't worry about it.

When the production toured to India, director Simon McBurney told
DNA Sunday that he had 're-shaped the scenes' and even 'thrown in
some Hindi'. McBurney, too, justified this with reference to an imagined authorial intention, implying another 'textual' approach towards
anachronistic disjunction: 'If Shakespeare was alive he would do so too'
(Jiwani 2005).

When stand-up comedians take Shakespearean roles, a modicum
of non-Shakespearean interpolation seems to be tolerated, but hardly
encouraged. When Alexei Sayle added the exclamation 'Bastards!' to
Trinculo's lines in a 1988 production of *The Tempest* at the Old Vic, the
Telegraph noted that Sayle 'interpolated some distinctly un-Shakespearean
language' (13 October); *City Limits* critic Carl Miller accused Sayle of
'head-butting his way through those of Trinculo's lines he actually keeps
in', though Sayle himself informed me that he only made the one textual alteration. Sean Hughes' Touchstone was also allowed the liberty of
one or two extratextual interpolations in David Lan's 2005 *As You Like It*.
As Touchstone was improvising his 'false gallop of verses' (3.2.97–112),
Hughes pretended to stumble on the line 'He that sweetest rose will find',
searching aloud for a rhyme with the words 'Rose ... petal ... stem ...
photosynthesis ...' before settling, once again, on 'Rosalind'. He also
mocked theatrical convention in 3.3, when the supposedly unseen and
unheard Jaques' asides were twice followed by the line 'Did you hear
that?' and finally by 'I definitely heard *that!*' I am not sure, however, that
the deviations from the text quite lived up to David Lan's suggestion in
his interview for the production's programme:

> There's a clown in the play so it seemed a good idea to find someone
> who had the experience of modern day clowning, working with an
> audience, working off the audience and improvising, which is basi-
> cally stand-up comedy, isn't it? And as Sean Hughes has great exper-
> tise in that field he was an ideal choice.
>
> (Lan 2005)

At no point was there ever any real sense of improvisation or audience
interaction in Hughes' performance. When another well known comedian, Ken Dodd, played Malvolio at the Liverpool Playhouse in 1971, he
was instructed strictly not to improvise at all and made only one ad lib –
a cover for a broken prop – during the whole run (Jackson 1996: 206).
Like Dodd, Sayle and Hughes might seem to have been muzzled by the
unspoken expectation in mainstream Shakespeare that deviations from
Shakespeare's script will be minimal and unobtrusive.

The further one looks from mainstream Shakespeare, however, the more interpolation one finds. In 1992, when Complicite were still considered an experimental fringe company and had not tackled a Shakespearean text before, their additions to the text of *The Winter's Tale* were much more manifest than their interpolations in the recent *Measure for Measure*. The production's co-director Annie Castledine explained to me that since Marcello Magni, the Italian actor playing Autolycus, had a very strong accent, she had suggested that he should say some of his lines in Italian and then explain them afterwards in contemporary English. Thus, Autolycus' 'three-pile' (4.3.14) became 'an Armani suit' (Holland 1997: 125) and his wares turned from Elizabethan ballads to bootlegged Stevie Wonder cassettes. 'Picking lecherously on women in the audience,' explains Holland, 'Magni tapped into a full tradition of *commedia* styles, playing out the timeless *lazzi* of trickery' (1997: 125). The production's heteroglossia was deliberate; as we saw in the first chapter, its purpose was to make it impossible for an audience to sustain any one response to the show (Castledine 2006).

In York-based company Riding Lights' production of the play in 2006, Tom Peters' Autolycus made similar use of anachronism. Improvising freely, he entered the stage with a karaoke set and introduced himself as a 'Bohemian street performer', before launching into a series of club-style gags and one-liners, promising 'a bit of blue for the Dads', and creating a definite disjunction with the self-reflexive joke: 'Shakespeare goes into a pub. The landlord shouts at him, "Get out, you're Bard!"' He also broke the 'fourth wall' for the first time in the play, bringing Sarah Finch's Dorcas onstage as his 'assistant' to hold up the lyrics to *Yesterday* and demand audience participation. But as we have seen, such freedom with the text breaks one of the widely accepted conventions of contemporary Shakespearean performance, and when I saw it in Canterbury, Riding Lights' production fell victim to this. As Peters' Autolycus entered the auditorium, an elderly gentleman on an aisle seat snapped crossly, 'Get on with the play!' The moment created a distinct tension between the staged jollity of the cast members and the resistance of certain members of the audience, heightened by the fact that Peters (an actor, not a stand-up comedian) chose to ignore the heckle rather than find a way of incorporating it into his act. As such, the audience response to Peters' act was, on this occasion at least, somewhat muted, and its effect diminished.

While the cases above are certainly examples of disjunctive anachronism, however, they are arguably once again 'surrogating' a primarily textual function: in allowing Autolycus to perpetuate the Shakespearean tradition of the anachronistic and satirical clown explored earlier in this chapter, they could be said, in a way, to be complying quite precisely with the design of the texts. Of course in this sense, this becomes quite inseparable from a 'metatextual' disjunction; the very act of jettisoning sections of Shakespeare's text becomes at once both a continuation of, and a separation from, the Shakespearean tradition.

In some productions, extratextual anachronism is pushed so far that it becomes the defining feature of the show, and as such, difficult to define as 'interpolation', since there is not even a nominal historical setting or a semblance of textual cohesion for it to disrupt. This is not to say, however, that such productions lose their disjunctive force; far from it. In the production of *Romeo and Juliet* by the Icelandic company Vesturport hosted at the Young Vic in 2003 (Figure 2.1), many of the lines were delivered in Icelandic; the first few lines of the prologue were spoken in this way, for example, before Víkingur Kristjánsson's Peter broke from the Icelandic script, grinned, and said, in English:

> Look at your English faces! *(audience laughter)* You're all going 'Good Lord, Cynthia, they're going to do it in Icelandic!' *(laughter)* 'Let's get out of here!' *(laughter)*

This opening set the tone for the rest of the production, in which no part of Shakespeare's text was safe from the company's irreverent treatment. Kristjánsson continued to interact with the audience, flirting with both male and female audience members throughout the play. Friar Laurence's cell was signified by an actor in a loincloth adopting a crucifix position; the same Christ-figure was later seen sharing a joint with the Friar. Audience members were asked to reach under their seats during Romeo and Juliet's wedding to find a bubble-blowing kit. Thor Kristinsson, meanwhile, played Paris as a schmaltzy variety singer, crooning songs like 'Oh, What a Beautiful Morning' and 'I've Got You Under My Skin' every time he appeared on stage. The production almost functioned as a variety show, interspersing elements of circus, stand-up, song, parody, and occasionally seriousness, to create an exhilarating and occasionally bewildering effect and a constant process of disengagement and reengagement with the Shakespearean text.

As we saw earlier, though, if too much is seen to have been 'imposed' upon a text, then a production can, in the eyes of the theatrical establishment at least, be seen to forfeit its claim to the name 'Shakespeare'. John Peter said of Vesturport's show in the *Sunday Times* that while it had 'the innate fire and energy' of Shakespeare's play, and while it may

Figure 2.1 Víkingur Kristjánsson as Peter in Vesturport's *Romeo and Juliet* (2003), directed by Gísli Örn Gardarsson.

have been 'a spoof, a romp, a game, even a loving homage ... one thing it isn't is Shakespeare' (*Sunday Times*, 12 October 2003). The *Sunday Telegraph*'s John Gross was more damning:

> The object of the evening is fun, and certainly almost everyone around me seemed to be having a good time. What they weren't getting was anything very much to do with the spirit or quality of the original play. ... The standard line of defence is that at least a show like this wins young people over to the general idea of Shakespeare, that it is a first step towards appreciating him properly. Perhaps; though I'm not sure that for lots of them it won't be an introduction to a lifetime of not appreciating him, accepting lightweight substitutes without being able to tell the difference.
>
> (*Sunday Telegraph*, 12 October 2003).

What these critics missed, however – or perhaps deliberately overlooked – is that the production was almost certainly never intended to be received as a faithful rendering of Shakespeare's play; Gross does not seem to consider that the object of the production might not have been to 'win young people over' to Shakespeare, or to encourage a 'proper' appreciation (whatever that means). This unapologetically irreverent production rather staged a *confrontation* with Shakespeare's cultural authority, putting into question the very assumptions implied in Gross' and Peter's reviews.

Even this production, however, was seen in some quarters as having 'surrogated' various aspects of the text. Michael Dobson's account of the show in *Shakespeare Survey* characterises it as one which was 'responsive to the play's own youthful excitement and spirit of playful improvisation', and draws attention to its use of aerial performance as 'making wonderfully literal the play's own interest in the imagery of up and down' (2004: 285–6). For Dobson, the production's iconoclasm was, paradoxically, also an act of textual fidelity: 'at once a funny, inventive, playful elaboration from the play and a very faithful and surprisingly upsetting translation of it into a rich, popular idiom' (2004: 286). Dobson's choice of words is telling: this was both an 'elaboration' and a 'translation'. Clearly we have begun to touch upon an issue here which exceeds the subject of anachronism. Extratextual interpolation in Shakespearean performance not only raises questions about the perceived sanctity of Shakespeare's script, but also forces an examination of what it is, exactly, that we mean by 'Shakespeare' in the first place; the further productions depart from Shakespeare's words, the more the line between 'interpretation' and 'adaptation' begins to blur (where, for example, do 'elaborations' and 'translations' fit in?). The next chapter addresses one aspect of this debate; it might, however, be usefully considered as an extension of this chapter's discussion of 'metatextual anachronism'. Chapter 4, meanwhile, approaches the subject from the opposite direction, beginning with a discussion of Shakespearean parody before meeting the kinds of production discussed here at the end.

Personal Narrative 3
Jeffrey Archer: The One that Got Away

I'm sitting in the Courtyard Theatre, Stratford-upon-Avon, midway through a daylong marathon of theatre. I took my seat for Michael Boyd's RSC *Henry VI Part I – The War Against France* at 10.30 this morning; I returned for *Part II – England's Fall* at 3 p.m., and by the time *Part III* finishes at around 11 p.m. this evening, I will have spent approximately ten hours of my day watching Shakespeare's trilogy. Several hours of visual spectacle, densely metaphorical staging and almost uninterrupted Shakespearean verse have already lulled me into an odd, dreamlike state.

This suddenly changes just over halfway through *Part II*. As the first of the 'Jack Cade' scenes begins (4.2), there is an immediate shift in tone. The lights are raised on the audience, shaking us from our reverie. Actors Forbes Masson and Jonathan Slinger, familiar by this point in the trilogy for their roles as the comic villains Alençon and the Bastard in *Part I*, enter from the back of the auditorium as the rebels Holland and Bevis. These roles are conceived as nightmarish clowns – both actors wear baggy, cartoonish smocks and thick black lipstick. They interact with the audience, using (for the first time in the trilogy) non-Shakespearean interpolations: 'Look at my axe!' bellows Masson on his way in, brandishing his weapon at a playgoer. The audience laugh, somewhat nervously: for Slinger and Masson are accompanied by a hooded captive who appears, by his pale contemporary suit and the programme in his hand, to be a member of the audience himself.

The tension mounts as the murderous clowns leave their captive onstage and return to the auditorium; they now approach a second audience member, a middle-aged man in a stalls aisle seat, and goad him onto the stage. They engage in some interplay with him, handing him the axe and indicating that he is meant to execute his hooded

56

fellow-playgoer, who has by this point been forced into a kneeling position. The would-be executioner plays along, making as if to swing the axe. The audience laugh again, but Slinger intervenes, narrowing his eyes. '*We* do the jokes!' he growls.

The clowns appear to have brought the man's bag onstage with him, and proceed to empty it out before his eyes, glaring confrontationally at him as they do so. Sundry items including a mobile phone, a wallet, and a paperback book are tipped onto the floor. Glancing at the book's title, Slinger reads aloud, '*Richard Three* ... What's that about, then?'

'I've read it,' replies Masson. 'It's shit.'

Another laugh passes through the audience: many are probably aware that the same company will shortly stage this play, with Slinger as Richard.

The clowns rifle through the man's wallet. Something in it apparently reveals him to be a magistrate, and they round on him. They take out what appear to be several ten-pound notes, and hand them to a third audience member, grinning nastily at their victim. Behind me, I hear some gasps, presumably from spectators incredulous at these actors' cheek. Then the captive's hood is lifted, and some of this tension is released: the man is recognisably a member of the company, and the real audience member is allowed to return to his seat.

Now, in a burst of movement and energy, the stage fills up with Jack Cade's followers. The fake audience member remains onstage for the scene which follows, but despite his twentieth-century suit, he does not look out of place: Cade's followers include other clowns dressed similarly to Masson and Slinger, a handful of actors in what can only be described as fancy dress (one wears a pair of angel wings), and even a couple of Pythonesque 'cripples' – a blind woman frantically taps her stick, while a woman with no legs pushes herself around on a wheeled cart. Some performers hang from the ceiling on ropes, performing aerial stunts. The emphasis is clearly on the carnivalesque, and on the grotesque body: comic depictions of disease, disability, and madness abound. Even death is subverted: some of the mob are played by the actors of main characters now deceased in the trilogy – the Cardinal, Gloucester, Talbot, Young Talbot – and though they are patently different characters here, they are still dressed in their original costumes.

The scene is filled with overlapping moments of anachronism and audience interaction. Dick the Butcher roars the famous line, 'The first thing we do, let's kill all the lawyers!' (4.2.78), and the whole company run to the edges of the stage to ask as many audience members as they can, 'Are you a lawyer?' After a flurry of interaction, the cry comes up,

'We found one!', and the rebels flock towards the unfortunate playgoer. No sooner has one such moment passed than another takes its place. The fake audience member still clutches the programme he entered the stage with, and one of the rebels uses it as proof of his bourgeois affiliations: 'He's a book in his pocket with red letters in't!' he cries (4.2.91; indeed, the programme was fronted with large red letters). The audience laugh as the man is sentenced to death, conspicuously aware of the identical programmes which many of them hold in their own hands. He is hanged by the mob, and his 'corpse' dangles above centre stage for the rest of the scene. Cade decides to 'make myself a knight presently' (4.2.118), and as he kneels and knights himself, he sparks off a chain reaction among his followers. They 'knight' first each other, and then the front row of the audience. 'What's your name?' says one such rebel to a teenage girl in the row in front of me; she replies, cowering and giggling, and is instructed, 'Rise up, Lady Katie!' In an emphatically Bakhtinian moment, as Cade confronts Sir Humphrey Stafford, the hanging corpse comes back to life and joins in with the riotous celebrations.

The sequence ends, and the transgressive energy subdues as we return to a scene between Henry and Margaret. I enter a more analytical frame of mind. I have just witnessed a truly carnivalesque sequence, in the most Bakhtinian sense: full of comic deaths and disfigurements, of anachronisms, of social role reversals; here, quite literally, no 'footlights' had separated actor and spectator (Bakhtin 1984: 7). The main plot was subjected to debasing mockery: Cade's spurious claim to royal descent is a brutal and anarchic parody of York's claim to the throne, no less bloody, and serves to throw into question the very concept of the divine right of kings and of the importance of royal lineage, showing York's claims (and by extension, those of his sons Edward IV and Richard III) to be mere political constructions. By implication, the same logic also applies to Henry VI's divine right, or indeed that of any monarch. If the Cade scenes were this anarchic in Shakespeare's time, I reason, their effect must have been highly subversive. But what, I find myself wondering, is the effect of such carnivalising for audiences here, today, at the Courtyard Theatre, Stratford-upon-Avon?

It passes from my mind as once again my concentration returns to the production. But talking with a member of the Front of House staff afterwards, the question returns to me. The audience member's bag, I learn, was in fact planted, although the audience member himself was genuine (the planted bag has, apparently, led frequently over the play's run to audience members worrying that they might have found a bomb).

Slinger and Masson generally pick on whichever member of the audience happens to be sitting in the seat by the bag: thus whoever sits in that seat, regardless of their real profession, becomes the 'magistrate' referred to in the clown's exchange:

HOLLAND.   The nobility think scorn to go in leather aprons.

BEVIS.   Nay more, the King's Council are no good workmen.

HOLLAND.   True; and yet it is said 'Labour in thy vocation'; which is as much to say as 'Let the magistrates be labouring men'; and therefore should we be magistrates.

(4.2.13–20)

This sets me thinking. It strikes me that it is eminently believable that the clowns might pick a magistrate at random from the RSC's audience – indeed the sequence derives some of its 'edge' from this very fact. It seems unremarkable, too, that later in the same scene, the anti-bourgeois rebels might also find a lawyer in the same crowd. We, the Courtyard Theatre audience, holding our Complete Works programmes and laughing at the RSC in-jokes, represent the same ruling-class intelligentsia that the Cade rebellion is depicted as seeking to overthrow.

My Front of House source relates one further anecdote, and suddenly the sequence has a genuinely contemporary resonance – though not a politically unambiguous one. One night, I learn, disgraced politician Jeffrey Archer had been sitting in the fated aisle seat by the bag. Naturally, the cast were very excited about getting him up on stage to play the example of the 'no-good' ruling class. Sadly, however, Archer moved – or was moved – during the interval.

He is, I note on my way home, listed in the *Henry VI* trilogy programme as an RSC Patron.

3
'A Play Extempore': Interpolation, Improvisation, and Unofficial Speech

> QUESTION. Mr Secretary, how do you respond to people who are saying that the fact that Omar and bin Laden remain at large and their whereabouts, the United States apparently having no clues as to their whereabouts is making – is beginning to make the United States look ineffective and at a loss?
> RUMSFELD. Well, you know, some things are neither good nor bad, but thinking makes it so, I suppose, as Shakespeare said. I just don't happen to think that.
> Pentagon briefing, CNN, 3 January 2002

Shakespearean scripture

Thus former US Secretary of Defence Donald Rumsfeld somewhat disingenuously (mis)quoted *Hamlet* in order to lend his equivocation the weight of Shakespeare's moral and cultural authority. Indeed, it is common practice for writers and speechmakers to resort to Shakespearean quotation as a means of bolstering their arguments; 'as Shakespeare said' is second only to biblical quotation as a rhetorical touchstone for affirmations of continuity with traditional moral values. *The Complete Works of Shakespeare* and the Bible: the two sources guaranteed to top the list of contributors to any good dictionary of quotations;[1] the two books which 'castaways' on BBC Radio 4's *Desert Island Discs* are assumed to want with them on their desert island as a matter of course. Shakespeare is cited as scripture just as one might quote the gospel: such adages as 'The better part of valour is discretion' (*1 Henry IV*, 5.4.118–19), 'Poor and content is rich, and rich enough' (*Othello*, 3.3.176), 'Neither a borrower or a lender be' (*Hamlet*, 1.3.75) and 'to thine own self be true' (*Hamlet*, 1.3.78) are examples which might readily spring to mind. Books compiling and sometimes analysing such nuggets of Shakespeare's 'wisdom' are widely available (Peter Dawkins' *Wisdom of Shakespeare* series, published in association with Shakespeare's Globe and with forewords by Mark Rylance, is one which appears to have had a significant effect on practice at the Globe).[2]

But just as, apparently, 'the devil can cite Scripture for his purpose' (*The Merchant of Venice*, 1.3.97), so can the erroneous cite Shakespeare for theirs. It is noticeable that most quotation dictionaries will ascribe Shakespearean passages such as those mentioned above not to the fictional characters who utter them within the plays, but to Shakespeare himself. Thus it becomes easy to lose sight of the fact that the line 'The better part of valour is discretion' is spoken by Falstaff in a sequence of comic cowardice, or that 'Poor and content is rich, and rich enough' are the words of Iago, a conniving villain engaged in a ruthless campaign of manipulation. Both 'Neither a borrower or a lender be' and 'to thine own self be true' form part of a long series of hackneyed truisms spoken by Polonius, a character later described by Hamlet as a 'foolish prating knave' (3.4.189); in *Shakespeare's Proverbial Language*, R. W. Dent suggests that 'every idea in the speech is a commonplace', counting no less than 14 possible references to well known proverbs (xxvi; 27–8). Donald Rumsfeld's reference was to a line from Hamlet's deliberately evasive and abstruse conversation with Rosencrantz and Guildenstern in 2.2, and the familiar line from *The Merchant of Venice* I quoted myself belongs to a scene in which Antonio, the character who utters it, later promises to continue to 'spurn', 'spit on', and call 'dog', the same character with whom he has implicitly compared the devil. In short, none of the Shakespearean quotations cited in this chapter so far can, or should, be unproblematically identified as an insight of 'Shakespeare's'.

Shakespeare and improvisation: A textual approach

A similar blurring of Shakespeare with his fictional creations is in evidence in much discussion of extratextual interpolation. Shakespeare's position on the matter of ad-libbing and improvisation seems to have been unambiguous, if a large number of Shakespearean critics are to be believed; Hamlet, in his advice to the players, says in no uncertain terms:

> And let those that play your clowns speak no more than is set down for them; for there be of them that will themselves laugh to set on

some quantity of barren spectators to laugh too, though in the mean time some necessary question of the play be then to be considered. That's villainous, and shows a most pitiful ambition in the fool that uses it.

(3.2.38–45)

For some of the most august names in nineteenth- and early twentieth-century Shakespearean criticism, the implication of this passage was that Shakespeare himself was attacking the practice of these clowns. Many critics identified a personal attack on Will Kemp (who had probably left Shakespeare's company shortly before *Hamlet* was first performed): Brinsley Nicholson argued as much in a paper on the subject in 1880 (57–66), while John Payne Collier suggested in his edition of the play that it would 'not be surprising if, besides laying down a general axiom as to the abuse introduced by the performers of the parts of clowns, Shakespeare had designed a particular allusion to Kemp' (Furness 1877: 230). John Dover Wilson surmised in his 1936 edition that, since there is no clown in the company of actors addressed by Hamlet within the play, 'these words seem directed against a real clown recognisable by Shakespeare's audience' (1936: 196); for George Ian Duthie, meanwhile, it was 'quite clear ... that in putting these strictures on the Clown into Hamlet's mouth Shakespeare must be expressing his own views, presumably with a real clown or clowns in mind' (1941: 233).

The idea has achieved almost continuous currency since then, not only in mainstream Shakespearean scholarship, but also in writings by theatre reviewers, actors, and directors. Anne Righter's influential *Shakespeare and the Idea of the Play* used the passage as evidence that 'Shakespeare may have been troubled by comedians who elaborated their own parts to the detriment of the rest of the play' (1962: 163). In *Shakespeare and the Uses of Comedy*, J. A. Bryant heard behind Hamlet's words 'an aesthetically sensitive author's pained awareness of what it means to hand over an intricately wrought contrivance to minds that may only partially comprehend it' (1986: 9). Following Nicholson and Collier, American actress and director Margaret Webster's *Shakespeare Without Tears: A Modern Guide for Directors, Actors and Playgoers* interpreted the passage as a 'public rebuke' which might have led to Kemp's departure (2000: 43). In what might be seen as a fairly typical use of the passage in theatre criticism, Arthur Holmberg remarked with some dismay that the extratextual interpolations in the Flying Karamazov Brothers' 1983 production of *The Comedy of Errors* were 'violating Hamlet's judicious advice to the players' (1983: 53).

The passage has also provoked the indignation of various proponents of theatrical improvisation. Comedian Tony Allen has suggested that if (as he suspects) Hamlet's comments were a thinly veiled public warning to Kemp's successor Armin, then Shakespeare was 'censuring the early development of stand-up comedy' (2002: 56–7). Radical writer and director Charles Marowitz is even more scathing:

> As for Shakespeare's audience predilections, they are squarely with the upper-class snobs whom he was constantly flattering in his maudlin dedications. Unscripted jokes 'though it make the unskilful laugh' (that is to say, the groundlings, the workers, the hearty proletarian masses on whom the popularity of his theatre depends) would make 'the judicious grieve' – that is, the toffs, burghers, aristocrats and generals for whom he has dished up 'his caviar'. ... God forbid there should be a moment of spontaneous improvisation which reveals the brilliance of a comic imagination to the detriment of the text. But if an actor *does* raise a laugh when 'some necessary question of the play is then to be considered', surely the playwright is at fault for providing the potentiality for comedy in a moment intended to be utterly serious?
>
> (1991: 41)

Marowitz's attack on 'Hamlet's advice' may, of course, be utterly justified, but it must be stressed that the Prince's advice is no more an indication of Shakespeare's opinion than any of the other Shakespearean passages cited above. For all we know, Shakespeare desired his clowns to adhere strictly to his scripts no more than he wanted his audience to 'kill all the lawyers' (*2 Henry VI*, 4.2.78).

David Wiles points out that the assumption that Hamlet is Shakespeare's mouthpiece is based upon a fundamental misunderstanding of the play's theatrical effects. The 'groundlings', he argues – dismissed by Hamlet only lines earlier as 'capable of nothing but inexplicable dumb shows and noise' (3.2.12–3) – 'would not have tolerated Shakespeare's play had they not been able to make the obvious distinction between "Hamlet" and Shakespeare' (2005: ix). These references to the tastes of the 'unskilful' are *jokes* – the workings of which (though contemporary performance is clearly by no means an infallible guide to Elizabethan theatre practice) were evident when Mark Rylance performed the lines at the reconstructed Shakespeare's Globe in 2000. Told by a playful Rylance/Hamlet (and the distinction between performer and character is always in flux in same-light conditions) that they were 'capable of nothing but inexplicable dumb shows', the groundlings let out a great cheer. When Rylance added 'and noise', the line provoked a burst of self-mocking laughter. As Peter Thomson has argued, 'it would be misplaced reverence to suppose that Shakespeare resented an audience's demand for instant pleasure. In all his plays, there are written-in opportunities for actors to invite responses from the floor' (2002: 139).

If today we are unable to perceive the distinction between Hamlet's stated instructions for performance and the performance style implied by Shakespeare himself, it may be because Hamlet's advice very closely resembles the central concerns of naturalism. At stake here is the very idea of Shakespeare's text as untouchable; as a form of 'scripture'. Elevated to the status of 'high art', Shakespeare's plays are constructed more often as *literary* phenomena than as blueprints for performance; Charles Lamb famously wrote in 1811 that he could not help 'being of opinion that the plays of Shakspeare are less calculated for performance on a stage, than those of almost any other dramatist whatever' (1963: 23). Following Lamb, critics such as Bryant make the argument that though the theatre is 'undoubtedly the best place to realise the fullness of a great many plays',

> it is by no means the best place to exhaust the meaning of a play that can legitimately be called a dramatic poem. The full experience of any poetry worth remembering will be solitary, not communal. So it is with the dramatic poetry of Shakespeare, regardless of the fact that his dramatic poems came to the world first as theatrical performances.
>
> (1986: 9)

But this dissociation of the Shakespearean canon from its theatrical roots is no natural progression; it is rather the result of a long process of the constant reinvention and renegotiation of Shakespeare's cultural standing.[3] It was only posthumously that Shakespeare's work was awarded the status of 'dramatic poetry'; during his lifetime, Shakespeare showed little interest in publishing the majority of his plays, implying that he was writing not for posterity, nor for solitary study, but for theatrical effect. Lanier explains that today the Shakespearean text is the site of an 'enormous investment' by high culture as the 'authentic' Shakespeare:

> Since it is treated as secular scripture, meaningful in its every detail, that text requires professionals to provide the specialised knowledge needed to plumb its ever-receding depths. Various practices

contribute to this view of the Shakespearian text: the editing and re-editing of Shakespeare; the academic practice of close reading, in which paraphrase is heresy; and the modes of prosodic analysis typical of British training for actors, exemplified by John Barton and Cicely Berry. The value accorded the Shakespearian text is so widespread that it seems perverse to think otherwise; where else might one locate the authentic Shakespeare than in Shakespeare's exact words?

(2002: 57–8)

But Lanier suggests that 'there are reasons to think that the authority invested in Shakespeare's language is misplaced'. He argues that, paradoxically, 'the concern to preserve its archaic features end up fostering an impression of antiquity at the expense of Shakespeare's currency, one of the effects after which he strove' (2002: 58).

As we saw in the first two chapters, Shakespeare's plays explore the tensions and troubled relationships between the contemporary and the historical aspects of performance, between the poetic and the vernacular, the sacred and the profane. Certainly passages of the plays may be profitably analysed as 'dramatic poetry'. But there are equally many passages which on the page elicit no 'insights', nor contain any inherent 'beauty'; which serve, in other words, only a *theatrical* function. Obscure clown sequences are all but impenetrable today precisely because they were once immediate and topical for their audiences. Privileging a theatrical approach over a literary one, then, it could be argued that in fact the only way such sequences can be 'faithfully' performed is, paradoxically, in a departure from Shakespeare's text.

Improvisation on the Elizabethan stage

Shakespeare was writing for a theatrical tradition not far removed from the pre-Renaissance forms of folk drama. Less than a decade before Shakespeare embarked upon his career as a playwright, the popular comedian Richard Tarlton was at the pinnacle of a career built upon doing exactly what Hamlet was later to condemn. Famed for his improvised responses to 'themes' suggested by members of his audience, Tarlton was remembered in Stowe's *Annales* of 1615 for his 'wondrous plentifull pleasant extemporall wit' (Chambers 1923: 2, 105). Tarlton died in 1588, but many (possibly apocryphal) examples of his 'extemporal wit' are documented in *Tarlton's Jests*, first published in 1600. These anecdotes describe Tarlton's antics in a wide variety of locations, from Queen Elizabeth's court to country inns, but most importantly for our case there are several examples cited from playhouses: one 'jest' relates the story of Tarlton's interaction with a playgoer at 'the Bull in Bishops-gate-street, where the queenes players oftentimes played'; another tells how Tarlton dealt with an audience member's mockery, 'he playing then at the Curtaine' (Halliwell 1844: 13, 16). Sometimes there appears to have been a discrete episode set aside for such improvisation after the play had finished – the anonymous narrator of *Tarlton's Jests* tells us, 'I remember I was once at a play in the country, where, as Tarlton's use was, the play being done, every one so pleased to throw up his theame' (Halliwell 1844: 28).

However, several related incidents clearly took place in the middle of a play. One story is of Tarlton's cutting response to a playgoer who 'threw an apple at him, which hit him on the cheek' as he was 'kneeling down to ask his father blessing' in a play (Halliwell 1844: 14). Another tells of how, during a performance of *The Famous Victories of Henry the Fifth*, Tarlton stepped into the role of the judge, taking 'a box on the eare ... which made the people laugh the more because it was he'. When Tarlton reappeared later in his regular role as the clown Derick, he made an ad-libbed reference to the incident; informed by another character that 'Prince Henry hit the judge a terrible box on the eare', Tarlton replied that 'the report so terrifies me, that me thinkes the blow remaines still on my cheeke, that it burnes againe' (Halliwell 1844: 24–5). A third anec-dote is worth quoting in full, since it indicates that some extra-dramatic exchanges during plays could in fact be significantly extended:

It chanced that in the midst of a play, after long expectation for Tarlton, being much desired of the people, at length hee came forth, where, at his entrance, one in the gallerie pointed his finger at him, saying to a friend that had never seene him, that is he. Tarlton to make sport at the least occasion given him, and seeing the man point with the finger, he in love againe held up two fingers. The captious fellow, jealous of his wife, for he was married, and because a player did it, took the matter more hainously, and asked him why he made hornes at him. No, quoth Tarlton, they be fingers:

For there is no man, which in love to me,
Lends me one finger, but he shall have three.

No, no, says the fellow, you gave me the hornes. True, sayes Tarlton, for my fingers are tipt with nailes, which are like hornes, and I must make a shew of that which you are sure of. This matter grew so, that the more he meddled the more it was for his disgrace; wherefore the standers by counselled him to depart, both hee and his hornes, lest his cause grew desperate. So the poore fellow, plucking his hat over his eyes, went his wayes.

(Halliwell 1844: 14–15)

It should be noted that recorded here are no less than five responses and counter-responses; the line '[t]his matter grew so, that the more he meddled the more it was for his disgrace' implies there may have been even more. Peter Davison draws particular attention to the line, 'to make sport at the least occasion given him', pointing out that it 'suggests that Tarlton eagerly sought such opportunities' (1982a: 38). The accuracy of these particular examples is naturally open to question, but it is significant that they were at least thought indicative of Tarlton's general performance style; the narrator describes the *Famous Victories* incident as 'no marvell, for he had many of these' (Halliwell 1844: 25).

Of course, Tarlton's improvisation bordered on the illegal. Since 1574, innkeepers had been forbidden by the Mayor and Common Council of London to 'suffer to be interlaced, added, mingled or uttered in any such play, interlude, comedy, tragedy or show any other matter than such as shall be first perused and allowed' (Chambers 1923, 4: 274). In 1581, Edmund Tilney was granted a royal commission which required 'all and every player or players ... to present and recite [their plays] before our said servant, or his sufficient deputy' (Chambers 1923, 1: 322). As Wiles suggests, Tarlton's status as the Queen's favourite probably afforded him some measure of licence (2005: 14–5), but the fact remains that any unscripted public performance in late Elizabethan society was potentially subversive, and must have generated in its audience at the very least that frisson which the unofficial and unsanctioned act will inevitably generate in a tightly regulated cultural sphere.

Will Kemp, the performer who played many of Shakespeare's first great clown roles, was also famed for his improvisations. Described by Nashe as 'that most Comicall and conceited Caualeire Monsieur du Kempe, Jest-monger and Vice-gerent generall to the Ghost of Dicke Tarlton' in the dedication of *An Almond for a Parrat* in 1590 (Chambers 1923, 2: 325), Kemp was, in many ways, Tarlton's successor. Little hard evidence exists to indicate the exact nature of Kemp's theatrical improvisations, but the 1607 play *The Travailes of the Three English Brothers* depicts him as willing to engage in extempore performance with a company of Italian players ('if they well invent any extemporal merriment, I'll put out the small sack of wit I ha' left in venture with them'),[4] and Richard Brome's 1638 play *The Antipodes* invokes both Kemp and Tarlton as

negative examples when a clown (Biplay) is criticised for holding 'inter-
loquutions with the Audients':

> LETOY. But you Sir are incorrigible, and
> Take licence to your selfe, to adde unto
> Your parts, your owne free fancy; and sometimes
> To alter, or diminish what the writer
> With care and skill compos'd: and when you are
> To speake to your coactors in the Scene,
> You hold interloquutions with the Audients.
>
> BIPLAY. That is a way my Lord has bin allow'd
> On elder stages to move mirth and laughter.
>
> LETOY. Yes in the dayes of Tarlton and Kempe,
> Before the stage was purg'd from barbarisme,
> And brought to the perfection it now shines with.

(Gurr 2002: 256)

Of course, these fictional presentations of Kemp cannot be unquestion-
ingly assumed to represent anything reliably true about the man himself –
as Wiles notes, 'censorship regulations and the evidence of reported
texts must oblige us to treat Brome's evidence as cautiously as we treated
Hamlet's' (2005: 35) – but they do at least indicate something of the
popular perception of Kemp, which lasted until long after his death.

Equally speculative, although perhaps more useful for our purposes,
are the conclusions which might be drawn from Kemp's self-contained
episode in the 1592 play *A Knack to Know a Knave*. The play was advertised as follows:

> A most pleasant and merie new Comedie, Intituled, A Knacke to knowe
> a Knaue. Newlie set foorth, as it hath sundrie tymes bene played by
> ED. ALLEN and his Companie. With KEMPS applauded Merrimentes
> of the men of Goteham, in receiuing the King into Goteham.

Kemp's episode of the 'applauded merriments of the men of Goteham',
however – despite its billing in the play's title – is barely two pages long
in the text as it has come down to us. Louis B. Wright drew the conclusion that 'Without doubt, the bulk of the clownery was omitted in the
printed version, or left for the improvisation of Kemp and his clowns'
(1926: 519), while Peter Thomson observes that '[i]n such knockabout

scenes, the discipline of the cue sheet might be legitimately displaced by the spontaneity of improvisation, with tempo and duration determined by audience response' (2002: 142).

Tarlton's and Kemp's unauthorised interpolations were by no means the only unscripted lines spoken on the Shakespearean stage. In her 1930 study of *The Extra-Dramatic Moment in Elizabethan Plays before 1616*, Doris Fenton found stage directions such as *The Trial of Chivalry*'s

Enter Forester, *missing the other taken away, speak anything, and exit*

and, in Thomas Heywood's *Edward IV*,

Iockie *is led to whipping ouer the stage, speaking some wordes, but of no importance.*

Fenton also identified at least 26 further plays in which the indication 'etc.' could be found within the scripted speech (Fenton 1930: 17–19). As to the number of instances in which players departed from their scripts *without* a textual indication that they should do so, of course, we can only hazard a wild guess. However, a reference in Philip Powell's *Commonplace Book* to 'one Kendal in a stage play in Bristoll' who 'spake extempore ... in dispraise of the noble Brittans' (Davison 1982a: 39) indicates that extra-dramatic improvisation was practiced as late as the 1630s, and this is corroborated by an entry in the records of Sir Henry Herbert, then Master of the Revels:

> Upon a second petition of the players to the High Commission court, wherein they did me right in my care to purge their plays of all offense, my lords Grace of Canterbury bestowed many words upon me, and discharged me of any blame and laid the whole fault of their play, called *The Magnetic Lady*, upon the players. This happened the 24 of October, 1633, at Lambeth. In their first petition they would have excused themselves on me and the poet.
>
> (Adams 1917: 21–2)

Clearly in laying 'the whole fault of their play ... upon the players' rather than upon the poet (Jonson) or indeed upon Herbert himself, the implication of this passage is that the players had departed from the officially sanctioned script, and that their interpolations had caused offence.

One further aspect of the *Hamlet* passage must be explored in relation to this point, since it might suggest a significant disparity between what was written and what was performed. When the First ('Bad') Quarto of *Hamlet* was published in 1603 – probably a pirated, half-remembered version of the text recorded by the actor who had played Marcellus, Valtemand, and Lucianus[5] – Hamlet's condemnation of extemporising clowns was followed by an additional, much more specific attack evident neither in the Second Quarto nor in the Folio version of the play. The punctuation here is based on Wiles's (2005: viii), which I think clarifies the performative implications of the speech:[6]

> And then you have some again that keeps one suit of jests, as a man is known by one suit of apparel, and gentlemen quote his jests down in their tables, before they come to the play, as thus:
>
> 'Cannot you stay till I eat my porridge?'
> and:
> 'You owe me a quarter's wages!'
> and:
> 'My coat wants a cullison!'
> and:
> 'Your beer is sour!'
> and blabbering with his lips:
> '...'
> and thus:
> '...'
>
> – keeping in his cinquepace of jests, when, God knows, the warm clown cannot make a jest unless by chance, as the blind man catcheth a hare. Masters, tell him of it.

The origin of this additional passage is, in Duthie's words, an 'apparently insoluble' problem (1941: 231). John Dover Wilson identified a possible source for two of the phrases in the 'cinquepace of jests' in *Tarlton's Jests* (Wilson 1918: 240–1), arguing in his edition of the play that in any case, 'this addition must be a personal attack upon a particular clown' (1936: 197). Certainly the specificity of the four phrases implies reference to a set of existing catchphrases. However, it is unclear whether this was a scripted part of the scene which was later cut, or an actor's interpolation. T. J. B. Spencer argues in the Penguin edition of *Hamlet* that it is 'the sort of passage which would soon become out of date and would therefore be cut in later performances' (1980: 240), while Nicholson's suggestion was that it may have been scripted by Shakespeare following an argument with Kemp, and removed after he and Kemp were reconciled (1880: 57–66).[7] Chambers, on the other hand, argued that the speech 'can only be a theatrical interpolation' (Chambers 1930: 418–19).

Davison, meanwhile, has pointed out that while the passage must 'surely' have been an ad-lib, it would, when spoken, have been indistinguishable from the body of the scripted speech. Since an ad-lib, he argues, 'is, of its nature, designed for the audience and intended to be recognised as such by them', Hamlet's interpolation, in being unrecognisable by the audience as an ad lib, must therefore 'have been a company "in-joke" ... directed at Shakespeare.' This, he argues, 'suggests very strongly that Shakespeare was anxious to control his clowns' (1982a: 41–2).[8] I am not convinced, however, that the passage was indistinguishable from the preceding text at all. As Weimann argues, Hamlet's full speech about clowns – comprised of both the generally accepted 'Shakespearean' passage, and the First Quarto addition – brings together and juxtaposes 'two different orders of authority' (2000: 23). In the 'Shakespearean' section, Hamlet the character adopts a regulatory tone, aligning himself firmly with the forces of censorship. In the interpolation, however, he elaborates unnecessarily; paradoxically, as Weimann points out, he 'is telling the players what not to do ... by doing it himself' (2000: 23). Here, Hamlet – or perhaps the actor playing him – is at the very least quoting from, if not engaging directly in, the 'ancient, almost ubiquitous practice of unscripted, unsanctioned performance' (2000: 24).

Shakespeare's scripted improvisation

This brings us to an important point regarding improvised performance. As Davison's analysis suggests, it is never entirely possible, as an audience member, to be sure whether a performer is truly improvising, or whether the material has been rehearsed in some way. Indeed, Keith Johnstone, founder of the improvisational theatre form Theatresports, explains in his book *Impro for Storytellers* that he has on occasion been accused of 'presenting rehearsed scenes under the guise of improvisations'; spectators, he claims, tend to 'believe even the worst scenes to have been rehearsed' (1999: 193, 25). But as Johnstone implies, it is not, ultimately, the lack of script which defines 'improvisational' speech. It is the order and nature of speech with which the performance aligns itself:

> The truth is that people come for a good time and nobody cares how the scenes are created except other improvisers. Dario Fo was entertaining seventy thousand people in a football stadium when lightning began ripping across the sky, so he launched into an impromptu debate with God. Was he improvising? Mightn't he

have been basing it on old material? Who cares? It must have been wonderful either way.

(1999: 25)

Problematic as it may be, if we attempt to look at Shakespeare's texts from the point of view of the Elizabethan audience, we can see that the plays make frequent use of those forms of speech – mostly prose, but also simple rhymes and verses – which would, to the ears of the playgoers, have *sounded* as improvised and as 'unofficial' as a genuine ad-lib. As Jane Freeman argues in her essay 'Shakespeare's Rhetorical Riffs', scripted drama and improvisation are not directly opposed to one another in Shakespeare's plays; rather, 'they function as parts of a continuum, for Shakespeare's scripted improvisation is both modelled on and a model of the extemporaneous dialogue of actual improvisation' (2003: 247).

Shakespeare scripts characters in the process of improvising throughout his plays, often in entirely self-contained episodes. In *The Winter's Tale*, Autolycus invents an elaborate set of lies in order to manipulate the Old Shepherd and his son (4.4.715–831); *As You Like It* features similar sequences of extended chicanery in Touchstone's trickery of both Audrey (3.3) and William (5.1); the gulling of Malvolio, particularly the 'Sir Topaz' sequence, is a particularly vindictive piece of improvised deception (*Twelfth Night*, 4.1); *The Merchant of Venice*, meanwhile, shows Launcelot Gobbo deceiving his blind old father for some length of time (2.2.30–105). All these sequences, it should be noted, bear at least a passing resemblance to many of the anecdotes related in *Tarlton's Jests*.[9] Shakespeare's greatest improviser, though, is of course Sir John Falstaff: like the characters listed above, Falstaff also engages in long sequences of comic deception, both as the would-be deceiver (his relation of the Gadshill incident in *1 Henry IV*, for example) and as the gull (as in most of *The Merry Wives of Windsor*). Significantly, it is also Falstaff whom Shakespeare scripts engaging in *theatrical* improvisation, calling for 'a play extempore' in *1 Henry IV* (2.5.282–3), and going on to assume the role of King Henry in the impromptu play which follows. A skilled improviser, Falstaff incorporates Mistress Quickly's interjections in three short bursts of extemporised blank verse (2.5.395–401) before enacting the rest of the scene-within-a-scene in prose. Falstaff's unconventional use of metaphor is also worth analysis, since as Wiles points out, the character typically 'embarks on a sentence without, it seems, himself knowing what the final object of comparison will turn out to be' (2005: 130); among Wiles' examples is the line, 'If I fought not with fifty of them, I am a bunch of radish' (2.5.186–7).

Prose is frequently employed by Shakespeare's improvising characters in marked contrast with the blank verse, which, as we have seen, was the language of official histories. Whereas verse is elaborate and almost impossible to tamper with without ruining the metre, prose is much closer to colloquial speech and indeed, as Rackin suggests, 'more difficult to memorise exactly'.[10] Falstaff's prose, she argues,

> with its seemingly inexhaustible capacity for wordplay and improvisation, vividly exemplifies this spontaneity and uncontrollability. It directly opposes the rigidity of official language and defies the regulation of prior restraint to which written texts were always subject in an age of censorship.
>
> (1991: 238)

Significantly for my argument, it does not really matter whether Kemp or whoever first played Falstaff was *really* improvising at all – the point is that the text did all it could to make it *seem* as though he was. Discussing the Gobbo sequence in *The Merchant of Venice*, Wiles points out that '[t]he fact that the clown speaks in prose when most of the play is in verse creates the *illusion* of spontaneity' (2005: 9; my emphasis). Illusory it may be, but the seemingly 'improvised' speech of the clown characters clearly aligns them with the unofficial, the unsanctioned, and the anachronistic.

Shakespeare made similar use of mock-improvised rhyme. Tarlton had been famous for his ability to respond to 'themes' in extemporised verse: Halliwell's edition of *Tarlton's Jests* documents at least 16 different instances. The book also describes Tarlton's early encounter with the young Robert Armin, who would go on to become Shakespeare's second clown: Armin is adopted by Tarlton as his protégé ('My adopted sonne therefore be, / To enjoy my clownes sute after me') and is depicted as falling 'in a league' with Tarlton's humour: we are told that

> private practise brought him to present playing, and at this houre performes the same, where, at the Globe on the Banks side men may see him.
>
> (1844: 23)

That Armin continued Tarlton's tradition of handling 'themes' is evident from Armin's own *Quips upon Questions*, a collection, as its title implies, of Armin's rhymed 'quips' in response to a variety of questions. One of them is a reworking of a verse from *Tarlton's Jests*.

Whenever Shakespeare's clowns and fools are scripted improvising in rhyme, then, this tradition is surely also being invoked. Naturally the device is particularly evident in many of the roles Armin is thought to have played – *King Lear*'s Fool lapses frequently into rhymed responses to other characters' lines, as (on occasion) does Feste in *Twelfth Night*. Touchstone's parody of Orlando's love poetry in *As You Like It* clearly resembles the Tarltonian tradition:

> TOUCHSTONE.   I'll rhyme you so eight years together, dinners, and suppers, and sleeping-hours excepted. It is the right butter-women's rank to market.
>
> ROSALIND.   Out, fool.
>
> TOUCHSTONE.   For a taste:
> If a hart do lack a hind,
> Let him seek out Rosalind.
> If the cat will after kind,
> So, be sure, will Rosalind.
> Winter garments must be lined,
> So must slender Rosalind.
> They that reap must sheaf and bind,
> Then to cart with Rosalind.
> 'Sweetest nut hath sourest rind',
> Such a nut is Rosalind.
> He that sweetest rose will find
> Must find love's prick, and Rosalind.
>
> (3.2.94–110)

But Shakespeare's reference to this clowning tradition was not limited to Armin's roles. Wiles identifies Launcelot Gobbo's rhyme, '[t]here will come a Christian by / Will be worth a Jewes eye' (2.5.41–2) as having 'the clumsiness of an improvisation' (2005: 9), while Davison points out that Armado and Mote's rhymed exchanges in *Love's Labour's Lost* (3.1.81–119) 'may have given the appearance of spontaneity' and could have been easily expanded upon in performance (1982a: 40).

This last point is worth elaborating upon before we move on to an analysis of improvisation in contemporary performance. This is admittedly highly speculative, but given the theatrical context described above – and the way in which Shakespeare himself made use of that tradition in his scripts – one might wonder whether Shakespeare scripted in opportunities for Kemp, Armin, and the like to make genuine use of their improvisatory talents within his plays. Scattered references to 'extempore' performance in Shakespeare's own plays show that an improvised play was at least a plausible prospect for his audience – we have seen Falstaff's and Cleopatra's already, and Polonius' reference to the Players' pre-eminence for both 'the law of writ and the liberty' (*Hamlet*, 2.2.402–3) has been interpreted as an allusion to both scripted and unscripted performance. Furthermore, both Hamlet's tirade against clowns and Quince's assurance to Snug that he may do the lion's part 'extempore, for it is nothing but roaring' (*A Midsummer Night's Dream*, 1.2.64–5) imply that there may occasionally have been scope for extempore performance within scripted plays.

John Russell Brown has argued that 'read with the eyes of a clown, Shakespeare's plays offer abundant cues for business and improvisation' (1993: 93). Certainly many of the clown episodes in Shakespeare's plays are self-contained, and of little significance to the main plots. Wiles points out that all of the roles Kemp is thought to have played 'are structured in order to allow for at least one short scene in which he speaks directly to the audience'; this monologue, he notes,

> is normally placed at the end of a scene, and thus seems to provide a format within which the clown may extemporise without risk to the rhythm of the play or direction of the narrative.
>
> (2005: 107)

As we saw earlier, Kemp's sequence in *A Knack To Know A Knave* was very probably expanded upon in performance, given the disparity between the length of the scripted sequence and Kemp's prominence in the play's billing. The same, I would argue, might be said of the role of Peter in *Romeo and Juliet* – a role we know to have been played by Kemp, and one which also seems insufficiently prominent for a 'star' clown. Davison constructs an argument along these lines for Armin's probable roles in *Macbeth* and *Othello*, finding it 'inconceivable that after so long a wait an audience would be satisfied with the short routine given to the Porter; this is an act easily open to extension by a clown' (1982a: 43). Some roles, in fact, almost *imply* extratextual improvisation: Kemp's sequences with (presumably) a living animal in the scenes between Lance and his dog Crab in *The Two Gentlemen of Verona* must surely have necessitated some improvisation, since a performing dog will rarely behave entirely predictably. Indeed, there seems little point in scripting a self-contained episode of clowning featuring a live dog if a certain degree of controlled chaos is not the intention on some level.

For centuries, received critical opinion has been that Shakespeare's incorporation of clownish elements, sequences, and characters in his plays was merely a concession to popular taste; that audiences expected to see their favourite clowns, and Shakespeare begrudgingly obliged. The Porter speech, Coleridge once suggested, was probably 'written for the mob by some other hand, *perhaps* with Shakespeare's consent' (Hawkes 1969: 215; my emphasis). Ronald Bayne's comments on Will Kemp for the *Cambridge History of English Literature* in 1910 suggested that the *Knack to Know a Knave* sequence discussed above showed

> how inevitably the improvising clown, with his license to introduce his own additions, was a discordant and incalculable element in the play, and hindered the development of artistic drama.
>
> (Bayne 1910)

Such comments seem almost illogical in their elitism. Kemp's time with the Chamberlain's Men was the era in which Shakespeare wrote such plays as *A Midsummer Night's Dream, Much Ado About Nothing*, and both parts of *Henry IV*, creating the roles of Bottom, Dogberry, and Falstaff presumably with Kemp in mind; to suggest that Kemp *hindered* the development of this drama is absurd. Just as unreasonable is the assumption that Shakespeare simply accommodated the clowns, tolerating them but nothing more: Marlowe's combative prologue to *Tamburlaine*, promising to rid the stage of 'jigging veins of rhyming mother wits / And such conceits as clownage keeps in pay', indicates that poets could explicitly reject the clown and still meet with commercial success, and besides, Shakespeare was clearly not so very obligated to create roles for his clowns that he had to write one into, say, *Richard II* (a play written during Kemp's time with the company). As we have seen, the dramaturgical design of many of Shakespeare's texts makes positive *use* of the clowning tradition, finding a place for it in the interplay of official and unofficial registers which is, as we have seen, central to the dramatic functioning of the plays.

Contaminating Shakespeare: A metatextual approach

The several hundred pilgrims who pass each year through the doors of Holy Trinity Church, Stratford-upon-Avon, are confronted with a stark order of preservation when they read the inscription over Shakespeare's grave:

Good frend for Jesus sake forbeare,
To dig the dust encloased heare.
Bleste be the man that spares thes stones,
And curst be he that moves my bones.

It might be an analogy for the similar taboo which rests today around the man's plays: cursed be the actor that tampers with Shakespeare's words. We saw at the opening of this chapter that in many ways the canonical Shakespearean text has become a kind of secular scripture; indeed its words are so well known that the incorporation in contemporary performance even of non-canonical passages which might be legitimately described as 'Shakespearean' achieves a kind of sudden dislocation. Actor Michael Pennington describes the effect in his book *Hamlet: A User's Guide*; in John Barton's 1980 production at Stratford, Pennington 'secured licence' to reinstate the First Quarto passage into the advice to the players, 'on the reasonable supposition that no-one would have heard it before'. The audience, he recalls, 'lulled with familiar words, didn't half sit up' (2004: 25).

As we have seen, Shakespeare made extensive use of the conflicting speech registers of the Elizabethan stage: the palpably 'unofficial' modes of real or scripted improvisation (dangerous and uncensored at least by association, if not in actuality) contrasted with the much more markedly 'official' registers of elevated, *locus*-centred characters. Today, however, the whole Shakespearean canon has become distinctively 'official',[11] and even the comic sequences which would once have been identified with unofficial registers of speech have achieved official canonical status. The Shakespearean text *itself* has now replaced the official, elevated discourse of official histories it once used only as part of a dichotomy. Passages which once subverted that discourse have now, paradoxically, become a part of it.

The taboo against the incorporation of extratextual passages, then, is in a way somewhat nonsensical. On the one hand, we cannot depart from the text without appearing unfaithful to Shakespeare, hovering on the borderline of literary sacrilege; on the other, it might be argued, we cannot fulfil the textual function of anachronistic or 'unofficial' passages in performance without some departure from that text. The problem thus becomes irreconcilable, and a strictly 'textual' approach unsustainable. As we saw at the end of the last chapter, the 'contamination' of Shakespeare's text with alien, 'unofficial' words will inevitably lead, intentionally or otherwise, to the delivery of some kind of commentary on that text: in other words, what I have called elsewhere a 'metatext'. The following section looks at such an interplay of 'Shakespeare' and non-Shakespearean, 'unofficial' speech.

Improvisation today

We saw some of the effects of scripted interpolations at the end of the last chapter. The most markedly 'unofficial' register of speech in contemporary performance, however, must be that which is, or appears to be, improvised. While the impositions of censorship on the modern stage have been nowhere near the strictures which were placed upon Elizabethan performers, improvisation was still technically illegal in Britain until the Theatres Act of 1968. Prior to that, improvised performances had been, in Keith Johnstone's words, an 'impossible dream', since all public theatrical performances had to be sanctioned in advance by the Lord Chamberlain's Office. Still a relatively new cultural form in this country, then, improvisational theatre is by its nature a challenge to the widely assumed primacy of the dramatist. It subverts the conventional, comfortable, one-way relationship between audience and play, and destabilises its audience's expectations. Bim Mason explains the thrill of the unsanctioned, unofficial act in his book on street theatre, describing the 'very real, exciting sense of danger' instilled among the crowd when a high-status audience member is imitated to their face and the threat of offence being taken becomes apparent (1992: 184).

This element of risk is perhaps what lies behind the attraction of improvised performance. In his writings on the subject, Johnstone aligns theatrical improvisation more with sport than with what he calls 'show business' – it was Johnstone himself who coined the term 'theatresports'. 'People so love spontaneity,' he explains, 'that they were reacting as if watching a sporting event': like a sport, he argues, improvisation threatens the possibility of failure (1999: 22). Johnstone challenges the assumption that 'failure has no value' in the theatre, claiming that sport's 'tug-o'-war between success and failure' is the reason for its compulsiveness as a spectacle; its excitement 'is maximised when there's a fifty/fifty toss-up between triumph and disaster' (1999: 66–7). The thrill generated by a constant flirtation with the possibility of failure was well described by Laurence Olivier in his book *On Acting*; comparing the excitement of watching a 'truly great comedian' with that of watching a 'fine Hamlet', Olivier wrote that:

The comic lives dangerously, he is always on the razor's edge. One moment of mistiming and his audience will turn on him. They are sitting there, waiting. Laughing – and waiting.

(1986: 152)

Unlike conventional theatre audiences, a 'sporting' audience will not sit passively through theatrical failure; the production's inability to engage them will be manifest.

Brecht once described the 'sporting public' as 'the fairest and shrewdest audience in the world' (1977: 6); elsewhere, he spoke of his desire for a 'theatre full of experts, just as one has sporting arenas full of experts' (1977: 44). Like sport, improvisation implies a much more active engagement on the part of the audience than is conventional in the contemporary theatre. As we saw in the first chapter, the assumption that Brecht was uninterested in engaging an audience on an emotional level is a common misconception; and as Johnstone's description of theatresports' origins in pro-wrestling makes clear, sporting audiences are far from emotionless:

Wrestling was the only form of working-class theatre that I'd seen, and the exaltation among the spectators was something I longed for, but didn't get, from 'straight' theatre.

(1999: 1)

Their key difference from 'straight' theatre audiences, then, is not that sporting audiences are more 'detached'; rather, they are more *dynamically* involved, and it is this very engagement which not only enables, but actively compels them to think critically about the events unfolding before them. It is for this reason that politically concerned theatre practitioners like Brazilian director Augusto Boal use improvisational forms in their work. Like Brecht, Boal's priority is to encourage audiences to 'no longer assume a passive, expectant attitude, but instead a critical, comparative one' (2000: 149); popular audiences, as he explains in his book *The Theatre of the Oppressed*,

are interested in experimenting, in rehearsing, and they abhor the 'closed' spectacles. ... Contrary to the bourgeois code of manners, the people's code allows and encourages the spectator to ask questions, to dialogue, to participate.

(2000: 142)

Boal describes his central objective as 'the liberation of the spectator, on whom the theatre has imposed finished visions of the world'
(2000: 155).

In an essay on laughter, the comedian and director Jonathan Miller
suggests that joking is a means of alienating oneself from what he
describes as the 'automatic pilot' of everyday life (1988: 16). Normally,
he argues, we 'mediate our relationships with one another though a
series of categories and concepts which are sufficiently stable to enable
us to go about our business fairly successfully' (1988: 11); but a rigid
adherence to these categories and concepts results in social inflexibility –
what Boal might call a 'finished vision of the world'. What we need,
therefore, argues Miller,

> is some sort of sabbatical let-out in one part of the brain and one
> part of our competence to enable us to put things up for grabs; to
> reconsider categories and concepts so that we can redesign our rela-
> tionships to the physical world, to one another, and even to our own
> notion of what it is to have relationships.
>
> (1988: 11–2)

This process of unconventional thought is analogous to what
Johnstone describes as the 'lunatic thinking' behind good improvisation; for him, it 'is the difference between soaring about in a limitless
universe and being locked up in a grubby little room' (1999: 72). The
image, while somewhat hyperbolic, suggests a politics inherent in
the very form of improvisation which is disjunctive, creative, and
potentially radical.

What I hope is beginning to become clear here is the synergy
between improvisational theatre forms and the kind of carnivalesque,
anthropologically 'popular' performance described in Chapter 1.
When the improvisational form is used in conjunction with scripted
text, then, we might expect to see something approximating the
'tussle and tension' between *locus* and *platea* discussed in that chapter, or of the interplay of official and unofficial registers discussed
in Chapter 2. The ways in which this interplay might function in
modern Shakespearean performance will be analysed in a moment,
but Mason's discussion of the improvised metanarrative around
the scripted narrative in his own company's show *Hell Is Not So Hot*
might suggest an example of its functioning in popular theatre more
broadly. This show, he says,

operated on two levels – the group of actors who argued, interrupted and commented on the play, and the narrative itself. ... This meant that dropping out of character, either to improvise, comment on the play, or deal with an interruption, could be done within the context of the piece.

(1992: 119)

Mason likens this to Brecht's alienation, suggesting that '[e]motional scenes could be played for real and then made fun of' (1992: 119). Naturally this multivocal, multiple-register performance style, typical of popular performance, is potentially useful for modern Shakespeareans.

Shakespeare in Improvisation: Official vs. unofficial

An interplay between the 'official' register of Shakespearean text and the unofficial speech of improvisation is evident in much improvised comedy. 'Shakespeare' is, of course, a staple of the genre. Verse challenges have always been a central feature of theatresports and the various entertainment forms it has spawned, and Johnstone himself uses examples from Shakespeare to train improvisers in verse (1999: 246). Theatresports games such as 'Film and Theatre Styles' (popularised on the television show *Whose Line Is It Anyway?*) will frequently list 'Shakespeare' as one of many acting styles for the performers to improvise a given scene in, and some troupes have gone even further and created a whole game based around improvising a full twenty-minute 'Shakespearean' play. London's Comedy Store Players, particularly when playing at a 'Shakespearean' venue like the Globe, the Swan, or Regent's Park Open Air Theatre, will often improvise a mock-Shakespearean play based upon a title suggested by the audience (past examples have included *Titus Androgynous*, *The Merchant of Bognor*, and *The Slightly Wet Weather*). The same game is also frequently played by the spin-off company Paul Merton's Impro Chums. Sometimes a whole show is sold on this basis: the Barbican's 'Everybody's Shakespeare' festival in 1994 featured a performance called 'Improvised Shakespeare', while Chicago's iO Theater (formerly the ImprovOlympic) hosts a weekly show by the Improvised Shakespeare Company which promises 'fully improvised plays using the language and themes of William Shakespeare' (iO Theater 2007).

The connection to Shakespeare in these instances is, of course, merely nominal. As the Improvised Shakespeare Company's director Blaine Swen admits, 'we could just as easily be the Improvised Marlowe Company. But we probably wouldn't be as popular' (Sharbaugh 2006).

'Shakespeare' in these instances tends to signify simply a fairly general-ised mock-Elizabethan register, sometimes verse, but more often simply deliberately nonsensical and obscure prose; the following transcript from a 1991 episode of *Whose Line Is It Anyway?* is an example of the latter:[12]

> JOHN SESSIONS.   *(throwing himself jauntily to the floor, waving his forefinger around, and adopting a stereotyped RSC-style delivery)* Come, my lord, even to the greenwood we will go. Faith, like two *[indecipherable]* we will dance in the sky of my hope, and here is something incomprehensible: there are two doits that are yet coits upon the face of it.
> *(audience laughter)*
> And there thou seest it, dost not?
>
> ARCHIE HAHN.   Ay, my lord, I would follow thee. I would point my arrow into the direction. ... I would disappear into the grove of sycamores and whack away at mine own interests for thee.
>
> SESSIONS.   Ay.
> *(laughter)*
> But a sycamore is but a yew, and a yew is true to you, and indeed to myself, and indeed to the lady that becomes to't again.

Improvisational comedy's engagement with Shakespeare, as in the example above, tends to lie in the humour generated by a head-on collision and confrontation between the 'official' tones of pseudo-Shakespearean dialogue and the 'unofficial' speech of improvisation. Mock-Elizabethan language might be employed in contrast with incongruous subject matters and cultural references, as in the following example from Robin Williams' 1979 show *Reality ... What A Concept*. Single-handedly performing an improvised 'Shakespeare' play on the subject of Three Mile Island, Pennsylvania (the site of a nuclear accident in March 1979), Williams delivered this doggerel prologue in an exaggerated English accent:

> Entertainer time, when all is simply strange,
> And yet radiation causes chromosomes to rearrange;
> My father worked at yon atomic plant hither,
> And sends me forth with my arms thus withered.

Williams's deliberate incongruity between official and unofficial registers is perhaps best illustrated by his response to audience heckling a few moments later, when he cried, 'No! Assholes do vex me!' A similar tension was at play when Mark Rylance performed with the Comedy Store Players in March 2004, and regular player Jim Sweeney stopped the show after Rylance improvised a particularly obscene line. 'You can piss in my mouth if you want!' repeated Sweeney, in counterfeit outrage; 'The Artistic Director of Shakespeare's Globe, ladies and gentlemen!'

The longest and most complex 'Improvised Shakespeare' shows I have seen were both devised and hosted by the comedian and maverick theatre director Ken Campbell and produced in conjunction with The Sticking Place: *Shall We Shog?* at Shakespeare's Globe in April 2005 (Figure 3.1) and *In Pursuit of Cardenio* at the 2006 Edinburgh Fringe Festival. Both performances also exploited the tensions between official and unofficial speech registers, but in what was perhaps a more interesting and ambiguous way. *Shall We Shog?* was essentially a 'theatresports' event in which three teams representing London, Newcastle, and Liverpool competed in a variety of improvised Shakespeare-themed, performance challenges, often based on audience suggestions.[13] These included delivering a 'Hey Nonny Nonny' song on any subject chosen by the audience (bear baiting, in this case), improvising sonnets, and re-enacting

Figure 3.1 The Sticking Place – The London team at *Shall We Shog?*, Shakespeare's Globe Theatre, 2005. (From left to right, Mike Mears, Sean McCann, Pearl Marsland, Adam Meggido, Cornelius Booth, Lucinda Lloyd, and Josh Darcy).

(or rather improvising) the 'famous' scene from Shakespeare's lost play *Cardenio*. *In Pursuit of Cardenio* was a development of this concept, featuring many of the same performers, which was performed over seven nights in hour-long instalments by a rotating cast of around six to ten. Drawing on passages from *Don Quixote*, the idea was to 'reconstruct' Shakespeare's play, but much of the event consisted of tongue-in-cheek explorations of various 'Elizabethan' concepts, from iambic pentameter and the four humours, to more eccentric ideas like the fabled 'School of Night' and the ancient art of 'gastromancy' (an art which, in the words of critic Rachael Halliburton, 'essentially combines mysticism with farting'; *Time Out*, 10 August 2006). More a series of workshops than a set of polished performances, the areas of exploration in Campbell's show changed from night to night, constantly adopting new approaches and incorporating audiences' suggestions (and, in some cases, personal belongings).

What was most striking in both shows was the ease with which Campbell's performers – clearly a highly educated and intelligent group of actors – improvised authentic-sounding 'Shakespeare'. Campbell encouraged them to improvise in song and in verse, to quibble, and to digress into elaborate metaphorical passages – tasks which many of them performed very credibly, drawing on genuine classical references as well as anachronism and deliberate nonsense, and which might have seemed ready-prepared were it not for the inclusion of references to audience members and their suggestions. Unlike the examples above, this was 'mock-official' speech: convincing as the genuine article until the illusion was intermittently punctured by intrusions from the 'unofficial' (performer Keddy Sutton seamlessly integrated the line 'Now, by the Power of Greyskull' – a reference to the 1980s cartoon series *Masters of the Universe* – into one passage of mock-Elizabethan blank verse). In both shows, Campbell introduced the audience to the theatrical tradition of 'nubbing' – the recitation by an actor of a deliberately meaningless piece of blank verse in the eventuality of forgetting one's lines during the course of a Shakespearean play. The rules of nubbing, he explained, were simple:

- one must use the word 'nub' at the beginning of the passage, to alert fellow actors to the departure from the script;
- one must then digress into a series of nonsensical quibbles;
- one must use the word 'nub' again towards the end of the passage, to indicate an imminent return to the text;
- the concluding words of the nub must be 'Milford Haven'.[14]

Following this introduction, Campbell's selected performers then improvised pieces of convincingly 'Shakespearean' blank verse to these very specifications – much to the audiences' delight.

It goes without saying that these examples of improvised verse, while successful, had nothing of Shakespeare's intricate and complex poetry. But contemporary performance, I would contend, rarely communicates the intricacy and complexity of Shakespeare's poetry anyway: forgotten or garbled lines can pass unnoticed as such, and in adaptations such as John Barton's *Wars of the Roses* or the Cheek by Jowl *Cymbeline* (studied in Chapter 4), the 'join' between the actual Shakespearian lines and the modern additions can be difficult to spot. If 'Shakespeare' is so evidently forgeable in performance, the binary between Shakespeare's official speech and the unofficial speech of improvisation becomes open to question. An implicit challenge, then, is levelled by such theatrical forgery, not only to Shakespeare's cultural supremacy, but also to the hard-and-fast categories of official and unofficial speech which underpin it: Shakespeare's words are just words, and if in theatrical production they are reduced (as they so often are) to impenetrable sound objects, they become no more meaningful than improvised nonsense.

The deconstructive edge to *Shall We Shog?* was underscored by precisely the element of unpredictability discussed above. Emblems of the 'official' contemporary Shakespeare were highly visible: the reconstructed Globe itself is in many ways a kind of totem to Bardolatry, and its then Artistic Director, Mark Rylance was conspicuous at the foot of the stage throughout the night. But equally prominent were more anarchic forces: the improvising teams, Campbell himself, Josh Darcy's colourful 'Goader' – who would keep score by impaling pieces of fruit on a pole, which he would frequently use to poke the improvisers – and most notably, the comedian Chris Lynam. Lynam, infamous on the comedy circuit for the routine in which he would set a lighted roman candle between his buttocks and sing 'There's No Business Like Show Business', was on this occasion dressed in a pink tutu, wearing heavy, gothic eyeliner, and prone to sudden outbursts of violent movement. The respective 'official' and 'unofficial' constituent elements of *Shall We Shog?* collided memorably and spectacularly when Lynam rugby-tackled Rylance to the floor of the Globe stage during the interval.

The evening's *frisson* between official and unofficial was most interesting, however, in a pre-planned, non-comic moment. As part of a lead-in to a series of acting games based around Hamlet's 'advice to the players' speech, Rylance was invited on stage to deliver the speech itself. As always, Rylance spoke his lines with a tremulous musical quality, and not only were the audience quite happy to indulge this speech, but in their intense and focused silence there was a tangible sense that they were following it very closely indeed. When Rylance stumbled upon a line, however, and broke that contact, the audience erupted with laughter, and Rylance made a quick ad-lib as an apple was playfully thrown at him from the yard. It seemed to me that the moment demonstrated very precisely Olivier's point, quoted above, about a 'fine Hamlet' performing constantly 'on the razor's edge'.

Improvisation in Shakespeare: Text or metatext?

As the latter example suggests, a tension between official and contemporary unofficial speech within the structure of a Shakespearean performance can bring about some interesting effects. The central question, then, is whether these effects construct themselves as 'textual' (that is, 'surrogating' a textual function) or 'metatextual'. In practice, of course, the two are most often intertwined.

In support of the case for the 'textual' effects of improvisation, Johnstone identifies a strong affinity between Shakespeare and improvisation, peppering his books on the subject with examples from Shakespeare's plays. Clearly the textual examples he gives are not improvised; they are cited, generally, as examples of the kind of 'unbounded imagination' he wishes to encourage in improvisation:

> We suppress our spontaneous impulses, we censor our imaginations, we learn to present ourselves as 'ordinary'. ... If Shakespeare had been worried about establishing his sanity, he could never have written *Hamlet*, let alone *Titus Andronicus*.
>
> (1981: 84)

Johnstone likens the Clowns' routine with Caliban in *The Tempest* to an improvisation, for example, on the grounds that it confirms and then subverts audience expectations (1981: 139). As we have already seen, Shakespeare was writing for performance in which there was a strong element of risk and of disjunction, which was almost certainly designed to provoke creative thought ('lunatic thinking', in Johnstone's phrase) among its audience.

Shakespeare's imitation of unofficial speech is perhaps one of the first *platea*-like performance elements to disappear in much contemporary performance, where the emphasis is nearly always upon speaking the lines very clearly and accurately. Paradoxically, as we have seen, perhaps the most faithful manner of delivery for such passages is in a relatively free delivery of the text itself. Michael Gambon's portrayal of Sir John Falstaff in the National Theatre's productions of the *Henry IV* plays in 2005 might be a case in point; while most reviews of Gambon's performance were generally positive, nearly all of them complained that his mumbling made some of the lines difficult to hear. Charles Spencer's review, for example, objected that

> If Michael Gambon's monstrously fat, funny and unbearably poignant Falstaff has a fault it is that the disreputable knight's drink-slurred voice isn't always clearly audible, and the eloquent words really matter with this character.
>
> *(Daily Telegraph*, 6 May 2005)

Indeed, it would appear that Gambon frequently garbled his lines. In *Shakespeare Survey*, Michael Dobson notes that Gambon 'never quite remembered the lines in precisely the same way every night; even on his best behaviour at the press night, Part 2's resolution "I will turn diseases to commodity", 1.2.249–50, became something that sounded more like "there's nothing ... worthful ... but commodities"' (2006: 325). However, one might argue that Gambon's relative inarticulacy was a positive strength of the performance. Slurring and stumbling over his words, he made them *sound* improvised, and the fact that they were thrown away identified them as unambiguously 'unofficial' speech – and as we have seen, there is a strong argument that this is precisely the effect Falstaff's words are designed to have. As Dobson's review goes on to point out, Gambon's mumbles were 'always characterized by a shrewd and potentially quite hostile sense of Falstaff as a predatory fraud, forever adjusting his often squalid behaviour to secure Hal's attention' (2006: 325). Actor David Harewood, who played Hotspur in the productions, noted that Gambon's performance was different every night:

> People were just coming off stage in fits of laughter, or looking at each other like, what on earth? Other actors might find out what works and stick with it, but Gambon never did that. He always tried to find something new.
>
> (Brockes 2006: 18)

Such an approach, I would suggest, may actually be 'surrogating' Falstaff's textual function in the play – as a foil to the official world of the court, as an element of danger and unpredictability, as a first point of contact for the audience, and as a representative of all that is debased, subversive, and anachronistic. In a modern performance, textually speaking, the less Falstaff sounds as if he is speaking 'Shakespeare', the better. Peter Davison makes a similar point concerning the Shakespearean recordings made by the comedians Stanley Holloway and Frankie Howerd. Howerd, playing Lance, apparently 'acts the words as best he may in the legitimate tradition', while Holloway's recording as Bottom – though he adds 'virtually nothing except slight repetitions' – has a 'much freer effect'. Davison argues that Howerd should have been allowed to 'work them up with an audience in his own style', suggesting that such an approach would be 'truer to the spirit of the plays' (1982a: 47–8).[15]

Genuine ad-libbing (as opposed to scripted interpolation) tends to be found more often at the Globe than any other mainstream Shakespearean theatre today. This distinctive aspect of performance at the Globe is perhaps a requirement of the space itself – unplanned intrusions from the extra-dramatic world are a regular distraction there, be they pigeons, planes, or simply the presence of a very visible standing audience – and acknowledgement of these distractions is often entirely necessary. As Patrick Lennox, who played Snout in the 2002 season's *A Midsummer Night's Dream*, explains:

> There are occasional times when there is an extremely noisy Chinook helicopter coming over, we might all break and look up because it just has to be brought into the story. It's not necessarily supposed to be funny, but you can't just stumble ahead if the audience focus has turned to that noise. To acknowledge that the focus has shifted elsewhere is the right thing to do – and if you do then they love that mutual acknowledgement. ... It's perilous to divorce yourself from outside stimulus. Once you've done that you can turn back to the play. Even if the noise doesn't cease, it becomes void once it has been accepted in that way.
>
> (Ryan 2002)[16]

Thus, for example, when a helicopter passed overhead during a scene between Stephano and Trinculo in 2005's *The Tempest*, Mark Rylance and Edward Hogg barely wasted a second in breaking from the script to wave at it for 'rescue' from the island. The audience, enjoying the sudden metatextual disjunction, rewarded the improvisation with a round of applause. Later, Alex Hassell also acknowledged a plane which passed, somewhat serendipitously, upon Gonzalo's statement that there would be 'no use of metal' in his imagined commonwealth.

The Globe's 2005 production of *Pericles* went further, however, than simply responding to distractions. In an extratextual framing narrative device – an overt 'metatext', if you will – Patrice Naiambana's Gower, drawing upon the African tradition of the *griot*, narrated to the older Pericles the story of his life, using a predominantly contemporary idiom (until the moment Marina's supposed death was related, the younger and older Pericles were played by different actors). Both Gower and the older Pericles remained on stage almost continuously, and Naiambana would improvise with the audience throughout the linking narrative passages, making playful jibes at our perceived resistance to textual 'contamination' ('If you came here for art, you've come to the wrong place. At the Globe, we do lowbrow!') and to the multicultural diversity of the production's cast and performance styles:

> NAIAMBANA/GOWER. *(pointing to each cast member in turn)* A Hungarian, an Italian, a Nigerian. ... Where are you from, Jules?
>
> ACTOR. Yorkshire!

When this actor's response prompted a round of applause, Naiambana wondered aloud, apparently baffled, what was so special about Yorkshire. He also made several topical jokes, drawing out the parallels between the text and the audience's own contextual 'metatext' quite pointedly. Concerning Pericles' responsibility to give aid to Tarsus, he referred to one of the big political campaigns of the year: 'You got to give aid. How are we going to Make Poverty History if you don't give aid?' Only days after the New Orleans disaster, his joke about President Bush's lacklustre response was decidedly edgy: 'Some kings,' he noted with a nod to the audience, 'don't even give aid to their own people!'[17] Reviews such as Nicholas de Jongh's, criticising Naiambana's 'philistine Gower, who mints far too many of his own lines, and believes art belongs in museums not in the theatre' (*Evening Standard*, 3 June 2005), were, I felt, missing the point. The very act of allowing Naiambana so much freedom to improvise levelled a challenge at the 'untouchable' nature of Shakespeare's text, providing an important and satirical metanarrative which was not at all incompatible with what was elsewhere a total commitment to Shakespeare's story.

When I suggested, in an interview with the several-time Globe actor James Garnon, that roles such as *Macbeth*'s porter, Autolycus, and

Launcelot Gobbo seemed 'designed to be ad-libbed upon', Garnon replied that 'it would make a lot of sense, and should probably be done'. He reasoned that in contemporary performance, 'all those sequences now fall completely flat'. But when I asked him why he felt actors and directors were reluctant to take that step in performance, he replied by suggesting that they were in fact held back by audiences:

> Some people at the Globe do. It's a very fine line. They start sniffing, that you're saying stuff; it sort of pulls them up, and they go, 'What are you doing? Why are you not saying Shakespeare?'
>
> (Garnon 2006)

Certainly Garnon's analysis tallies with the kind of response I witnessed at Tom Peters' Autolycus in Riding Lights' production, described at the end of the last chapter.

But I wonder whether Globe audiences are as conservative as they might appear. A model for the potential of ad-libbing in Shakespeare in the theatre might be provided by Garnon's own performance in the Globe's *The Storm*, a verse play by Peter Oswald, based on a Plautine comedy. This was a production which was unafraid to make use of anachronism; Mark Rylance's character Daemones assured the audience early on that

> We've got permission. The playwright actually phoned up Plautus and said, Titus Maccius, is it all right if we're not strictly period? And Plautus answered, 'Look Pete, I wouldn't be talking to you now if it wasn't for anachronisms. You go ahead and use them!'
>
> (Oswald 2005: 12)

This disjunctively anachronistic approach extended to improvisation. In one sequence, Garnon would regularly try to make Rylance corpse – not, as might be assumed, an example of actors misbehaving, but (according to director Tim Carroll in a post-show talk) a genuine directorial decision.[18] In another sequence, the production advertised its flirtation with the unpredictable, with Garnon's character engaging in direct audience interaction:

SCEPARNIO.   Plesidippus! Plesidippus! He's either dead or out of ear-
    shot, in which case there's no way on earth I can reach him!
    Never has been! If a person's out of range of the sound of your
    voice, you can't speak to him, that's that, you can project and

enunciate until your brain bursts, you're basically talking to yourself! Oh if only a hand would reach out of the future with some as-yet undreamed-of device by means of which you can speak to a person who isn't there! Will such a miracle ever exist? Then I could save my sweetheart!

(Oswald 2005: 50)

Garnon would then labour the point until an audience member indicated that they had a mobile phone, and would proceed to converse with the relevant party. The sequence which followed was, I was told, always genuine improvisation; on the first show, apparently, England were playing an important cricket match, and Garnon was thrilled to discover a text message about it, which he read aloud. Whatever happened in between, though, the sequence inevitably culminated in Sceparnio's realisation that since he did not have Plesidippus' number, the phone was useless. Thus, in a way – much like the monologues Shakespeare scripted for his clowns – the sequence was self-contained and able to adopt a performance register somewhat different from the rest of the show.

Several companies have, of course, allowed those that played their clowns to 'speak more than is set down for them', and the results have, predictably, been both successful and contentious. We have already seen the press response to Vesturport's *Romeo and Juliet*, a production which made use of a large amount of extratextual ad-libbing and direct address (the Nurse, Friar Laurence, and Capulet all addressed the audience, and Peter did little but). Footsbarn's *A Midsummer Night's Dream*, performed in a big top in London's Highbury fields (among many other tour venues) in 1991, divided opinion with Mechanicals who spent much of their stage time ad-libbing ('Think brick. Imagine yourself as a baby wall', Cousin 1993: 28), and a Puck who communicated only in non-verbal, guttural sounds. The *Independent*'s Sarah Hemming, like many other press reviewers, praised the 'tremendous rapport with the audience' built up by the clowns' improvisations (*Independent*, 26 June 1991), but a number of other critics were less enthusiastic. John Peter, for example, accused the production of 'mutilating the author and patronising the audience', describing some of the departures from the script as 'barbaric' and others as 'simply pointless', and wondering why 'one of Shakespeare's most accessible plays needs to be adapted for popular consumption' (*Sunday Times*, 30 June 1991).

I saw a particularly successful example of improvisation 'surrogating' a textual function, while also providing a disjunctive metatext, in open-air company Illyria's performance of *The Tempest*. Actor Marcus Fernando' s first entrance as Trinculo, from the back of the audience, was entirely ad-libbed. Joking about the potential level of audience bewilderment as far as the plot was concerned, he found an audience member who was attempting to follow the text in an Arden edition, and flicked through it, wondering aloud whether Trinculo was to have 'any good love scenes'. After a great deal of interaction, involving the theft of several picnic snacks and two glasses of wine from audience members, he entered the scene with Caliban and Stephano, having rooted these characters very much in the unofficial, *platea*-like performance register. Both Stephano and Trinculo continued to make contemporary references and to ad-lib with audience members throughout the performance.

Naturally when actual improvisation starts to become a feature of performances, the argument that it is performing merely a 'textual' function – faithful, in Davison's phrase, 'to the spirit of the plays' – becomes more difficult to sustain (1982a: 47–8). Clearly an improvisation will always provide a 'metatext' of sorts. W. B. Worthen writes of a 1998 production of *Romeo and Juliet* at the University of California, Davis, in which a 'gifted comic actor with experience in stand-up comedy' was cast as Peter, Will Kemp's role; Worthen describes the performance as 'not an imitation of Kemp, but a surrogation of Kemp's function, using an embodied performance to mark a history outside the text, a history also traced – just barely – within it' (2003: 75). Such moments of surrogation, argues Worthen, 'are neither governed by the text, nor seem to restore "original" behaviour' (2003: 76). The next chapter will explore the potential effects of the more radical interplay between text and metatext which Worthen's analysis suggests.

Personal Narrative 4
A Bit Sexist

It was just a snatch of conversation, overheard during the interval in the gents' toilets at the Old Vic: the initial reflections of a young audience member upon the all-male company Propeller's 2007 production of *The Taming of the Shrew*.

'It's a bit sexist, innit?' he commented to his friend.
'*Obviously* it's sexist,' replied the friend, with some condescension.

The production had employed a very farcical mode of performance for the first few scenes, and there had been a strong sense that it was inviting us to laugh. As it went on, however, our laughter was interrogated and troubled in a variety of ways. The first half of Petruchio and Katherine's wedding scene made the most of every available opportunity for comedy: Dugald Bruce-Lockhart's Petruchio entered in a thong and a tasselled leather jacket, swigging beer and urinating into his cowboy hat. Following a break, however, in which the events of the ceremony itself, taking place offstage, were narrated by Gremio; the second half of the scene changed tone entirely. Petruchio's nastiness was inescapable, as he hurled Katherine and chairs alike across the stage, before violently dragging her off, clutching her face. Baptista's following line – 'Nay, let them go – a couple of quiet ones!' (3.3.112) – was met with awkward and patchy audience laughter, and Gremio's 'Went they not quickly, I should die with laughing,' did not sound remotely jocular. The cast were clearly doing their best to subdue the audience's laughter rather than foster it, adding a very real tension to the scene.

Our complicity in Katherine's pain was emphasised increasingly as the production continued after the interval. At the end of 4.1, Petruchio delivered his 'thus have I politicly begun my reign' speech in a downstage spotlight, speaking directly to the audience. After his demand,

> He that knows better how to tame a shrew,
> Now let him speak. 'Tis charity to show.

> (4.1.196–7)

an extratextual addition followed:

> *(to an audience member in the front stalls)* You wanna say something?
> *Audience laughter.*
> *(taking in whole audience)* Anyone want to say anything?
> *Uncomfortable pause.*
> *(with a shrug)* Okay …
> *Exit.*

Shortly afterwards, Katherine directed her curse on Grumio at the whole audience:

> Sorrow on thee and all the pack of you,
> That triumph thus upon my misery!

> (4.3.33–4)

The suggestion, of course, was that in our enjoyment of her suffering during the earlier, more farcical scenes, we had implicitly endorsed the very behaviour which had led to her present abjection.

This was most starkly emphasised in the play's final scene. Simon Scardifield played Katherine's infamous speech of submission very slowly, out to the audience, as if her spirit had been utterly broken; the fact that Katherine herself was being played by a male performer served only to emphasise that the values she was blankly parroting were distinctly patriarchal ones. Immediately following this quiet and under-stated moment, however, Petruchio's response – 'Why, there's a wench!' – was delivered loudly and confrontationally, like a bully who had just proven his point to a crowd of sceptics. This abrupt switch in register provoked an audience laugh; Bruce-Lockhart paused and stared at the audience until their laughter had died down, before adding, 'Come on and kiss me, Kate.'

The extent to which we had been complicit in enjoying the play's patriarchal wish-fulfilment was inescapable. It had been 'a bit sexist'. And it was obvious.

4
'It's like a Shakespeare play!': Parodic Appropriations of Shakespeare

IACHIMO. Hold, Sirrah!
What Briton slave is this?
POSTHUMUS. 'Tis Iachimo. Thank thee, gods.
They fight.
For me, my ransom's death.
Thou knowest me not Italian signore but I will spare
thy life.
Exit.
[...]
IACHIMO. What miracle is this?
An unknown Briton spares my life.

(Cheek by Jowl 2007: 80–1)

JOAN. Bloody complicated, innit? It's like a Shakespeare
play! Everyone's either miserable or dead! I rather wish
I'd stayed in the Costa del Sol!

(Rice & Grose 2007: 14)

Parodic appropriation

The practice of adapting, parodying, and otherwise appropriating Shakespeare is so incalculably widespread that this chapter on it must inevitably be selective in the extreme, and confine itself to one very specific aspect of the subject. The two quotations heading the chapter will, I hope, give some indication as to where this focus will lie. Both are extracts from mainstream adaptations of Shakespeare's *Cymbeline* staged in Britain in 2007 – by the theatre companies Cheek by Jowl and Kneehigh respectively. Both extracts are largely composed of newly written, non-Shakespearean dialogue (the only exception being the words 'For me, my ransom's death' in the Cheek by Jowl script). It is the difference in attitude displayed by the two towards their shared source which will concern us here. Where Cheek by Jowl's adaptation does its best to hide its disjunction from Shakespeare's script with deliberate use of archaic language and pseudo-Shakespearean asides, Kneehigh's playfully advertises the chasm separating it from its seventeenth-century forebear, making irreverent and mocking use of Shakespeare's play. In this respect, the attitudes of the adaptations towards their original might run parallel to the model of assimilative and disjunctive anachronisms laid out in Chapter 2.

Concerned as it is with popular appropriations of Shakespeare, this chapter will depart from the pattern established in the previous three; namely, that of a 'textual' or historical analysis followed by a consideration of 'metatextual' approaches. Any appropriation or parody is, in some sense, a 'metatext', since the practice involves maintaining a dual awareness at once of both the source text and the parodying text (see Rose 1979). Popular culture, as we have seen, can be characterised by its fostering of just such a double consciousness – and indeed by its opposition to, and disjunction from, 'elite' culture – and as a result, the focus of this chapter is largely on those appropriations of Shakespeare which might be considered parodic.

Like so much in the study of popular culture, it is impossible to generalise about the politics of parody, particularly when it stages a confrontation (as Shakespearean parody frequently does) between elite and popular forms. On the one hand, the irreverent appropriation of elite cultural elements by popular forms can be seen as an empowering, even subversive, act of transgression; on the other, parody can be culturally elitist in itself, appealing as it does only to those with enough knowledge of the parodied text to understand its references. In appropriating the 'highbrow', it might be argued, popular parody dissolves distinctions between high and popular art; but by another line of reasoning, parody is divisive in its derision, constructing a solid and impenetrable antagonism between elite and popular forms where perhaps the reality is more ambiguous.

French philosopher Henri Bergson famously propounded a conservative view of the function of laughter, finding in it 'an unavowed intention to humiliate, and consequently correct our neighbour' (1956: 148) and an impulse 'to suppress any separatist tendency' (174); by this definition, parody might be seen as a ridiculing force whose function is to police the boundaries between cultural categories. Mikhail Bakhtin, however – one of the great theorists of parody – took issue with Bergson's representation of laughter. During the Renaissance, he argued,

> the characteristic trait of laughter was precisely the recognition of its positive, regenerating, creative meaning. This clearly distinguishes it from the later theories of the philosophy of laughter, including Bergson's conception, which bring out mostly its negative functions.
>
> (1984: 71)

For Bakhtin, parody is an inherently carnivalesque and dialogic genre, poking fun at the 'straightforward word' of established forms, showing their discourses to be 'one-sided, bounded, incapable of exhausting the object', and their authority to be relative and contestable. Parodic laughter, he suggests, is 'a critique on the one-sided seriousness of the lofty direct word ... richer, more fundamental and most importantly *too contradictory and heteroglot* to be fit into a high and straightforward genre' (1981: 55).[1] Parody, in Bakhtin's formulation, is then dialogic, overlaying one set of discourses with another. 'Thus,' he argues, 'every parody is an intentional dialogised hybrid. Within it, languages and styles actively and mutually illuminate one another' (1981: 76). Carnival acts as a temporary suspension of cultural hierarchies.

In his discussion of Shakespearean appropriation, John Drakakis follows Bakhtin's dialogic model of parody. Parodic quotation of Shakespeare, he argues,

> systematically challenges monologic authority, serves as a means of giving the process of mythologising Shakespeare a specific historical context, and promotes the idea of Shakespearean texts as sites of con-testation, as opposed to being repositories of cultural wisdom.
>
> (1997: 165)

Drakakis uses the term 'demystification' to describe this function of Shakespearean parody, suggesting that its carnivalisation of the canon serves to deconstruct the reverence with which this icon of elite culture is treated, and allows some form of reappropriation by popular culture. Christy Desmet takes a similar line when she argues that 'Shakespearean appropriation contests bardolatry by demystifying the concept of author-ship' (Desmet & Sawyer 1999: 4). As we shall see when we turn to exam-ples of such pop culture appropriation, this is both true and untrue.

In the first chapter, we noted Robert Shaughnessy's speculation as to the potential avenues through which 'the critical insights of theory' might 'inform and transform our practical negotiations with Shakespeare' (2000: 13); the last two chapters have, I hope, suggested ways in which the addition in performance of various sorts of 'metatext' might begin to answer this question. In the parodies and adaptations studied in this chapter, we begin to encounter metatexts whose critical commentaries on Shakespeare are far more explicit. Any parody will inevitably provide a form of criticism upon its original; indeed, W. H. Auden once wrote that in his ideal 'College for Bards', '[t]he library would contain no books of literary criticism, and the only critical exercise required of students would be the writing of parodies' (1968: 77). As Simon Dentith points out in his study of parody, the genre 'forms part of a range of cultural practices, which allude, *with deliberate evaluative intonation*, to precursor texts' (2000: 6; my emphasis). Dentith suggests that parody differs from these other forms (and 'adaptation' might be considered among them) primarily on the grounds of the *inflection* it gives to its quotation. Parody, inherently double-voiced, cannot fail to take up an attitude towards its source in a highly conspicuous fashion.

What, then, might define the 'parodic inflection'? Criticism provides a range of answers to this question, many of them contradictory. A 'hypertext' must take up a 'playful' or 'ludic' attitude towards its source in order to be classified by Gérard Genette as parodic (1982: 453); Margaret Rose, meanwhile, describes the widespread understanding of parody as '*comic* quotation, imitation, or transformation' (1993: 6; my emphasis) and identifies 'the creation of comic incongruity or discrepancy' as a 'significant distinguishing factor in parody' (1993: 31). 'Comic' is a broad category, however, containing a range of different 'attitudes' – and as we shall see later in the chapter, critics such as Linda Hutcheon have found an 'insistence on the presence of comic effect' in parody too narrow anyway (1985: 20). However its inflection is defined, parody will always be on some level ambiguous; and a parody's 'attitude' is ultimately unquantifiable, lying as it does in the minds of its audience rather than in any essential or formal qualities.[2]

The range of Shakespearean parodies' attitudes towards their sources might broadly be said to lie somewhere on a spectrum between respectful homage on the one hand, and deliberate sacrilege on the other. As we have seen with such neat categorisations, however, reality tends to be a little less clear-cut, and this is particularly the case in discussions of Shakespeare's collisions with popular culture. Shakespeare's own relationship with cultural authority was always complicated, but now that Shakespeare himself has become a pop culture token (and often a fairly arbitrary token) of cultural authority, the matter is even more troublesome.

Many of the dialogic and carnivalesque performance modes associated with the puncturing irreverence of parody – clowning, innuendo, politically subversive jokes, audience complicity, self-reflexivity – were features of Shakespearean performance in the first place, and in some instances, Shakespeare's own carnivalesque sequences have been performed with all the anarchic disrespect of a parody, while not actually being 'parodied' in any literal sense at all. In any case, Shakespeare's texts are rarely the sole target of a parody (if indeed they are the target at all): the set of cultural symbols posited by a parody as 'Shakespearean' are very often simply the trappings of Shakespeare within a particular context, making reference to specific performance modes or the uses to which 'Shakespeare' is put at a given cultural moment. Shakespearean parody almost always pulls in two or more directions simultaneously. It is appropriation, certainly. But of what? For whom? And for what purposes?

The sections that follow will attempt, with necessary caution, to begin to categorise, with reference to examples, some of the different 'attitudes' which Shakespearean parody might take towards its source. I have drawn extensively not only from live performance, but also from television, since the theatre alone can give a fairly skewed sense of what 'Shakespeare' means within popular culture. I should also point out that though the history of pop culture's appropriation of Shakespeare extends back just as far as Shakespearean performance itself, I have confined myself to discussion of twentieth- and twenty-first-century appropriations: Shakespeare's meanings within popular culture have changed quite radically over the centuries, and there is, after all, a wealth of writing on pre-twentieth-century Shakespearean parodies and adaptations already.[3]

For initiates only

After mislaying the magical lamp at one point in the Old Vic's 2005 pantomime of *Aladdin*, Ian McKellen's Widow Tawnkey turned to the audience and confessed to being 'a very foolish, fond old woman'. For a pantomime audience, this *King Lear* reference was met with a surprisingly robust laugh.

Of course, this was not a typical 'pantomime audience', and if one were feeling uncharitable, one might argue that the joke did not earn its robust reception because it was particularly funny, but rather because it allowed the audience to demonstrate their credentials as a theatre-literate crowd. The production was peppered with such references: 'Now I am alone ...' said one character, taking in those audience members who laughed at the allusion to *Hamlet* with a playful sweep of his eyes; 'We'll be here longer than a Trevor Nunn production!' complained another, and the audience roared. These were appropriations of high culture, perhaps, and certainly they were being incorporated into a popular theatre form, but truly 'popular' appropriations they were not. They were jokes which rewarded knowledge of Shakespeare and the theatre world with a sense that one knew enough to belong to the in-crowd, those who 'got the joke', to the exclusion of the uninitiated. Linda Hutcheon discusses the 'pleasure of recognition' afforded by such parodic references (1985: 95), but notes that this kind of parody 'can work toward maintaining cultural continuity' (1985: 99); in drawing a line between the savvy and the ignorant in such a stark manner, the cultural politics of such moments are overwhelmingly conservative.

Similar (if rather more ambiguous) examples of such in-group joking might be found in *The Shakespeare Revue*, a compilation of Shakespeare-themed songs and sketches put together for the RSC by Christopher Luscombe and Malcolm McKee. The show was first performed in The Pit on 30 October 1994 as part of the Barbican's *Everybody's Shakespeare* festival; it transferred to the Vaudeville Theatre in London's West End the following year, when it was also recorded before an invited audience at Abbey Road Studios. Crammed with jokes about Adrian Noble and Trevor Nunn, its appeal was, as Jack Tinker pointed out, 'to insider traders' (*Daily Mail*, 24 November 1995); in a feature for *The Times*, Susan Elkin complained that 'most younger theatregoers will simply not understand the jokes' (11 September 1995), a claim which was lent further credence by her fellow *Times* correspondent Valerie Grove's 'Diary' some months later:

> having dragged two mid-teen daughters there, I had to concede that it really is more enjoyable if you are familiar with most of the plays; and I'm afraid they just aren't.
>
> (*Times*, 30 December 1995)

Neither writer indicated that this was a problem with the production itself, however. Elkin cites the 'cultural assumptions' made by the production as entirely reasonable ones and blames schoolteachers for the discrepancy in cultural knowledge between the older and younger members of the audience (her article is somewhat hysterically entitled 'The plot in schools to bury the Bard'). The production was implicitly pledging its allegiance to Shakespeare as a necessary component of cultural fluency, separating the culturally competent from the incompetent by means of shared laughter. Examples of the production's in-jokes for the Shakespeare-literate are plentiful; perhaps one of the funnier instances is from the sketch 'And How is Hamlet?' (an extract from Perry Pontac's verse radio play *Hamlet Part II*), in which a returning ambassador discovers what has happened in Denmark in his absence and enquires after the health of the inhabitants of the court:[4]

AMBASSADOR. The fair Ophelia?

FOURTH GENT. Foul Ophelia, sir.
　　　　For she lies decomposing, though her wits
　　　　Rotted before her.

SECOND GENT. 'Twas her father's death.

AMBASSADOR. The fair Polonius?

THIRD GENT. Dead, sir, dead as well.
　　　　Slain by Prince Hamlet who, as you have heard,
　　　　Is also dead.

AMBASSADOR. How tragic for the Queen.

The CD soundtrack records a delayed laugh just after this last line, as the audience, familiar with *Hamlet*, realise that Gertrude, of course, is dead at this point too.

The same writer was responsible for 'Othello in Earnest', one of the show's funniest and most popular sketches, but also one of its most culturally conservative. Ostensibly, the scene transplanted the characters of *Othello* into *The Importance of Being Earnest*, and 'Lady Brabantio' was witnessed interviewing Othello as a prospective husband for her daughter Desdemona in a direct parallel to the analogous scene in Wilde's play. It was primarily a very clever wordplay exercise, requiring of its audience a confident familiarity with *The Importance of Being Earnest*, but its puns had very little to do with *Othello* and everything to do with reinforcing some outmoded and uncomfortably racist African stereotypes. Thus, as Othello described his upbringing – 'my nappy a banana leaf, my rattle a quiver of poisoned arrows, my cradle a sandbag' – Lady Brabantio repeated the last two words with horror, to delighted applause from the audience. Upon his description of a lion attack, she protested, 'the lion is immaterial!' The sketch's punchline is perhaps the clearest example:

LADY BRABANTIO. Mr Othello, you seem if I am not mistaken to be showing signs of considerable self-esteem.

OTHELLO.　On the contrary, Lady Brabantio, I've now realised for the first time in my life the vital Importance of Being Burnish'd!

Elsewhere in the revue, there was a sense that innovation in Shakespearean performance was being censured. When director Deborah Warner cast Fiona Shaw as Richard II at the National Theatre in 1995, *The Shakespeare Revue* added a song which mocked cross-casting, provocatively entitled 'PC or not PC' (thus aligning cross-casting with a favourite *bête noire* of traditionalists, and assigning to it an ideology with which it arguably has no affinity). 'Now transvestite casting's becoming all the rage,' asked the song, 'What can we anticipate upon the London stage ...?' Its suggestions had more than a hint of Bergson's suppression of 'separatist tendencies' by means of ridicule:

Miss Suzman has mastered
Her Edmund the bastard,
Cordelia's Peter O'Toole;
And here's Jodie Foster,
With guide-dog, as Gloucester –
Lord knows who's playing the fool!

Of course, the cultural politics of this production was not entirely unambiguous. Michael Billington described the show as a 'sprightly anthology that assumes Shakespeare is not an elitist conspiracy but still part of our common culture' (*Guardian*, 14 November 1995), and many of the sketches it incorporated, borrowed from sources such as *Monty Python's Flying Circus* and *Beyond the Fringe*, were much more irreverent in tone. But ultimately, this was an evening in which Shakespeare was venerated through both mockery and homage, and the subversive impulses of some of the sketches were largely overwhelmed by the context in which they were performed. Robert Butler's review for the *Independent* probably best sums up this show's missed potential for a parodic interrogation of 'Shakespeare' the cultural icon:

Just as *The Day Today* caricatures the way the media presents the news, so this could take a really close look at the way the theatre, academia and the heritage industry bend Shakespeare to their own purposes. *The Shakespeare Revue* doesn't quite do this. ... It is not the vehicle for an attitude or point of view.

(*Independent*, 19 November 1995)

I would argue, though, that the show was a vehicle for a very specific 'attitude or point of view', but that this attitude was simply a conservative one. The inclusion, for example, of an entirely serious passage of J. B. Priestley's writings in which the gardens of Shakespeare's home in Stratford-upon-Avon were almost religiously invoked as 'the place where he was still alive' (Luscombe & McKee 1994: 66), left the audience in no doubt as to the attitude they were expected to take up towards 'the Bard'.

Of course, not all parodic references to Shakespeare requiring knowledge of his plays affirm such culturally conservative tendencies. There are certain tropes of Shakespearean performance (often, as we shall see, derived from Olivier's films) which have entered the popular consciousness through repeated popular allusion – Hamlet with Yorick's skull, Juliet on her balcony, and so on. The assumption that the majority of an audience would be familiar with the line 'To be or not to be, that is the question' is not an elitist one, for example, and does not encourage quite the same sort of self-congratulatory laughter as some of the allusions mentioned above; thus Monty Python's 'man who speaks in anagrams' can comfortably provoke a laugh with his quotation from *'Thamle'*, 'Be ot or bot ne ot, tath is the nestquoi'. A sketch about Shakespeare performed at a 1989 AIDS benefit by Rowan Atkinson and Hugh Laurie provides an example, in fact, which makes use of the audience's familiarity with the line in order to undermine their expectations. Atkinson plays Shakespeare's Blackadder-like agent, self-confessedly concerned only with 'bums on seats', and Hugh Laurie plays Shakespeare himself very much like the upper-class idiot George from the same series. At first, we are on familiar territory: the agent suggests trimming 'some of the dead wood' from *Hamlet*, and the natural assumption is that he is going to suggest some sort of heretical cut; the joke predicated, of course, on the familiar cultural stereotype that producers tend to be mercenary philistines, and on the shared value that Shakespeare wrote no 'dead wood'. The agent continues to play into expectations, describing the soliloquies as 'that stand-up stuff in the middle of the action', and when he begins to describe 'the dodgy one', the audience are led to believe they know where they are heading:

> AGENT. 'To be' … 'nobler in the mind' … 'mortal coil'. That one.
> It's boring, Bill.
> *Big audience laugh.*

Here, however, expectations start to be reversed; we learn that 'Bill' was forced to cut the 'avocado monologue' from *King Lear*, and then when we hear Shakespeare recite his speech, the sketch reaches its first big punchline:

SHAKESPEARE.　'To be a victim of all life's earthly woes, or not to be a coward and take death by his proffer'd hand.'

AGENT.　There, now I'm sure we can cut that down.

When the agent suggests 'To be or not to be', his outraged client replies, 'You can't say that! It's gibberish!' Finally, the agent suggests a compromise: 'All right, I'll do you a deal: I'll trim this speech, and you can put back in those awful cockney gravediggers.' Here, the audience's knowledge of Shakespeare is used precisely to subvert his cultural authority; the sketch's depiction of Shakespeare as pretentious and overly verbose, and his money-minded agent as the true author of Hamlet's most famous line, undermines shared cultural assumptions in a mischievous and subversive fashion.

Confrontational parody

Much of the audience laughter aroused by the sketch discussed above drew its force, I would like to suggest, from an underlying hostility to the cultural authority represented by Shakespeare. Lines such as 'It's boring, Bill,' and 'It's gibberish!' provoked the kind of transgressive laughter which greets the articulation of the 'unsayable', and recall Freud's construction of laughter as 'tendentious' and a means of venting aggressive impulses which might otherwise remain unarticulated:

> [T]endentious jokes are especially favoured in order to make aggressiveness or criticism possible against persons in exalted positions who claim to exercise authority. The joke then represents a rebellion against that authority, a liberation from its pressure. The charm of caricatures lies in this simple factor: we laugh at them even if they are unsuccessful simply because we count rebellion against authority as a merit.
>
> (Freud 1976: 149)

The analysis might bring to mind another, much more literal confrontation between Blackadder and Shakespeare: in the short film *Blackadder Back and Forth* (filmed for the Millennium Dome in 1999 and finally broadcast on BBC1 in 2002), the time-travelling title character encounters William Shakespeare in the court of Elizabeth I, and knocks him to the floor with the words:

> That is for every schoolboy and schoolgirl for the next four hundred years! Have you any idea how much suffering you're going to cause? Hours spent at school desks trying to find *one* joke in *A Midsummer Night's Dream*! Years wearing stupid tights in school plays, and saying things like 'What ho, my lord!' and 'Oh look, here comes Othello, talking total crap as usual!'

The 'tendentious' energies of the passage are unmistakeable: through ridicule, the joker and the laugher 'achieve in a roundabout way the enjoyment of overcoming' their shared cultural enemy (Freud 1976: 147).

Naturally a parodic appropriation of Shakespeare which constructs him as the 'enemy' of popular audiences is not doing much to reappropriate him for popular culture. However, as Freud later notes, such joking is often aimed not only at people, but also at institutions which are perceived as commanding an overbearing level of authority (1976: 153). What may at first appear to be a hostile parody of 'Shakespeare' is very often staging a confrontation between popular culture and the elite cultural uses to which Shakespeare is put. Even the *Blackadder* passage cited above – deriving its comic effect from a physical assault upon Shakespeare himself – takes issue with Shakespeare's appropriation by the education system rather than the man himself or his plays. As Lanier notes in his discussion of 'Shakespop parody',

> [t]ypically, Shakespeare per se is not the object of critique. Rather, these appropriations target the sorts of social and interpretive decorum that govern how high art is treated, as well as those who enforce that decorum, authority figures like teachers, intellectuals, antiquarians, actors, and bluebloods.
>
> (Lanier 2002: 54)

Very often, it is the perceived pretentiousness of mainstream Shakespearean theatre practice which is sent up in such parody. Thus Stephen Fry and Hugh Laurie's sketch 'Shakespeare Masterclass' from their 1981 Cambridge Footlights revue *The Cellar Tapes* mocks John Barton–style textual analysis:[5]

FRY. Shakespeare's "T" is very much upper case, there, Hugh, isn't it? Why?

LAURIE.   'Cos it's the first word in the sentence.

FRY.   Well I think that's *partly* it.

(Luscombe & McKee 1994: 25)

Beyond the Fringe, of course, lampooned traditional 'Shakespearean' acting in their sketch 'So That's The Way You Like It'; a stage direction in *The Complete Beyond the Fringe* notes, '*This is played with great vigour at tremendous speed in the modern Shakespeare style*' (Bennett et al. 2003: 108). The BBC2 sketch show *Big Train* continued this tradition with a 1998 sketch in which a group of Shakespeare lookalikes, dressed in identical Elizabethan outfits, fondly remember Portaccio, the 'one true master' among Shakespeare lookalikes: the piece sends up the sonorous, rhythmic intonation, stilted body language, and histrionic acting style associated with pompous Shakespearean performance, down to every last raised eyebrow and affected chuckle. The sketch's similarity to some of the BBC Shakespeare productions of the early 1980s is marked.[6]

By far the most common allusion to established styles of Shakespearean acting is the 'Laurence Olivier impression', which has become so widespread in popular Shakespearean parody that it has perhaps become the default voice with which to signify 'establishment' Shakespeare. This can be seen, for example, in Christopher Reeve's *Muppet Show* Hamlet (discussed below), or as Drakakis has pointed out, in Tommy Cooper's sketch 'Tommy the Troubadour'.[7] Peter Cook's performance as Richard III in the very first episode of *Blackadder* ('The Foretelling') is a pastiche of Olivier's famous film characterisation, revealing his apparent hunchback to be the result simply of a trapped cloak as he delivers the lines:

Now is the summer of our sweet content
Made o'ercast winter by these Tudor clouds.

Olivier's *Henry V* is referenced in a similar manner, echoing the film's visual iconography as Richard leads his troops into battle with the words:

Once more unto the breach, dear friends, once more;
Consign their parts most private to a Rutland tree![8]

Perhaps the most famous parodic allusion to Olivier, however, is Peter Sellers' version of the Beatles' *Hard Day's Night*, which he recorded in 1965 (making the Top 20). Sellers caricatures the idiosyncrasies of Oliver's verse-speaking – accent, offbeat pauses, rhythmic delivery, sudden and dynamic changes in tempo and pitch – and in applying them to the lyrics of a popular song, he satirically deconstructs their aggrandising effect. Sellers performed a televised version of the piece in costume as Olivier's Richard III, as part of the broadcast *The Music of Lennon and McCartney* (transmitted 16 December 1965).

Incongruity

The central joke in the Sellers sketch, of course, is that he is delivering the 'lowbrow' in the manner of the 'highbrow'. This, or something similar, is the mechanism by which a huge amount of parodic appropriation of Shakespeare derives its comic effect. This might be said to work, broadly speaking, in one of two ways:

1. The plots or characters of Shakespeare's plays are transplanted into an alien setting or performed in an inappropriate style.
2. Alien or inappropriate content is performed 'in the style of Shakespeare'.

What is construed as 'alien' or 'inappropriate' is, of course, in most cases, the debased, the mundane, or the popular.

The tradition is a long one, dating back long before the twentieth-century remit of this chapter; it was the basic principle underlying Shakespearean burlesque from Thomas Duffet's 1674 parody *The Mock-Tempest; or, The Enchanted Island*, when Prospero's island became a prison for prostitutes, through to the countless (and very popular) travesties of the nineteenth century, which would rewrite Shakespeare's dialogue into crude rhyming verses and incorporate popular contemporary songs. Shakespeare's appropriation by music hall and variety theatre was similar, too: Worton David and Harry Fragson's song *The Music Hall Shakespeare* (c. 1905) set paraphrased speeches from *Hamlet*, *The Merchant of Venice*, and *All is True* to popular music hall songs (*Let's All Go Down the Strand*, *Oh! Oh! Antonio*, and *It's a Different Girl Again*), while in *I'd Like to Shake Shakespeare* (c. 1915), Mark Sheridan sang of his wife's 'passion for Shakespeare', relating the various quotations from his works which she would drop incongruously into their everyday domestic life. In a sketch in which he played a caricature Shakespeare with a 'ludicrously tall forehead', the music hall star George Robey would tell the story of going into a shop with a blank-verse request for cigarettes:[9]

> Hast thou among thy merchandise
> A brand of cigarettes thou canst commend?
> Not so full flavoured that their fumes
> Strike harsh upon my unsuspecting throat
> Nor yet so mild that they insipid be
> Like maid's first lisp of love beneath the moon,
> Nor yet unmindful of my slender purse
> That doth but ill provide my worldly wants.
> If such a brand thou hast I now beseech
> Let not thy action lag behind thy speech.

In its longer version, the speech continued in this manner for another 29 lines, before the punchline:

> And let me have 'em quick: I'm in a hurry.
>
> (Harding 1990: 160–1)

But variety theatre reached what was perhaps its most thorough appropriation of Shakespeare in the 1940s and 50s, with the acts of Leon 'Shakespeare' Cortez. Cortez would retell Shakespearean plots in a cockney dialect:

> Well, King Duncan and 'is sons Malcolm and Donalbain arrive, and bein' a bit tired the old cock goes straight for a kip, and no sooner 'as 'e dropped off to sleep than Mrs Mac, just like the cat, crept up to 'is cot, copped 'is clock, coughed and crept out again. 'Ain't yer done 'im?' sez Macbeth. 'No,' she sez – ''is clock reminded me of my father. You 'ave a bash.'
>
> (Wilmut 1985: 166)

Intriguingly, the British Empire Shakespeare Society once sent Cortez a letter congratulating him on his portrayal of the 'frailties and humanities' of Shakespeare's characters (Luscombe & McKee 1994: 93).

Today, the great majority of popular appropriations of Shakespeare work along similar lines, drawing comic capital from a perceived clash between Shakespeare and the popular or lowbrow. On the BBC sketch show *Dead Ringers*, for example, we saw the characters from *Hamlet* on the tabloid chat show *Trisha* (subtitle: 'My Mother Married The Uncle

Who Killed My Father!'); the League of Gentlemen's live show (2001) opened with football fans watching *Hamlet* and chanting 'Get thee to a nunnery!' and 'C'mon, Hamlet, make your fucking mind up!' At least three shows at 2006's Edinburgh Fringe Festival used the device: *Shakespeare for Breakfast* transplanted Shakespearean characters into reality TV show *I'm a Celebrity, Get Me Out of Here!*, *Macbeth Re-Arisen* was a blank verse sequel to Shakespeare's play drawing on popular motifs from the zombie film genre, and the much-hyped *Bouncy Castle Hamlet* was exactly what its title suggested (though sadly the concept ended there). Stand-up comedy seems to have created a subgenre, in which a supposedly inappropriate celebrity is impersonated performing passages from *Hamlet*: Jay Leno used to do an Elvis-as-Hamlet routine (c. 1974), Robin Williams delivered his impression of Jack Nicholson's 'To be, or not to goddamn be' in *Throbbing Python of Love* (1983), while more recently, Eddie Izzard's *Sexie* envisaged Christopher Walken in the role (2003).

Often the humour lies in an inappropriate coarseness of presentation. A 1978 episode of *The Muppet Show* featured Peter Sellers reciting the first few lines of *Richard III* while 'playing tuned chickens'; a 1980 episode showed Christopher Reeve (complete with Laurence Olivier-style costume and English accent) attempting to perform *Hamlet* while being undermined by Muppet intrusions. As always in such appropriations, though, there is an ambiguity as to where the humour is supposed to lie: at times, perhaps, we laugh with ridicule at the attempts of the Muppets and their guests to perform high culture; at others, we might enjoy their perceived transgression into it. The *Muppet Show* sketches are in some senses carnivalisations of 'highbrow' Shakespearean performance, the *Hamlet* scene in particular (with its talking 'Yorick' skull) making use of what Bakhtin termed the 'grotesque body' in order to 'degrade' a symbol of high culture. Such mockery draws attention, by means of slapstick and other forms of 'bodily' humour, to the physical and material, when the scene itself appears to demand a focus on the 'metaphysical' (though whether or not this is entirely true of 5.1 from *Hamlet*, opening as it does with its comic gravediggers, is of course debatable). It might be understood in similar terms to those in which Drakakis explains Tommy Cooper's appropriation of the same scene: 'a carnivalesque demystification from below of an episode which has always elicited reverential awe from Shakespearean critics' (1997: 168).

Carnival, however, is always only a temporary inversion, and when the theatre has extended a similar concept to full-length shows, it has encountered problems. Elly Brewer and Sandi Toksvig's *The Pocket*

Dream, which transferred to the West End in 1992 from a sold-out run at Nottingham Playhouse, was a play-about-a-play, in which a touring production of *A Midsummer Night's Dream* is almost cancelled when all but two of its cast choose to boycott it in favour of the pub; at the last minute, the stage manager, a stagehand, the front-of-house manager, and the leading lady's ex-boyfriend step into the roles, and farcical chaos ensues. Oddsocks Theatre, meanwhile, have carved a niche in 'Shakespeare pantomimes', with *Hamlet the Panto*, *King Lear the Panto*, and *Macbeth the Panto* among their credits. I saw the latter in Portsmouth in 2004, and found it full of traditional panto-style puns – 'It's a carrion crow!' 'What's it carryin'?' – and audience participation. Duncan, upon every one of his entrances, would ask the audience 'Who put the dunce in Dunsinane?', to which the audience were primed to reply, 'Duncan did!'

While both *The Pocket Dream* and *Macbeth the Panto* attracted their share of positive responses, one might argue that both were weakened by the extension over the course of a full-length play of what is essentially a single concept (that is, the performance of Shakespeare in a comically inappropriate style). Clive Hirschhorn of the *Sunday Express* criticised *The Pocket Dream* as an 'extended revue sketch' (8 March 1992), and this charge was widely repeated in other reviews; what might have made a successful sketch, it seemed, was difficult to sustain over two hours. Similarly, there were moments in *Macbeth the Panto* – Lady Macbeth's sleepwalking scene, Macduff's reaction to the slaughter of his family (and, in this show, favourite chicken), or Macbeth's nihilistic soliloquy in 5.5 – in which Oddsocks' relentless cheeriness sat uncomfortably at odds with the play's oppressively sinister content. I was reminded of Peter Brook's warning of the pitfalls of deliberately 'rough' theatre:

> Just as in life the wearing of old clothes can start as defiance and turn into a posture, so roughness can become an end in itself. The defiant popular theatre man can be so down-to-earth that he forbids his material to fly.
>
> (Brook 1990: 80)

Bisociation

One might argue that in most of the cases discussed above, a stress on the *incongruity* between Shakespeare and popular forms served to emphasise the cultural distance separating 'high' from 'low'. In this sense, these parodies were perhaps not only lending weight to the notion that Shakespeare is inaccessible to popular audiences (and thereby reinforcing cultural elitism), but in many cases also constructing the popular cultural forms from which they were drawing – popular songs, cockney dialect, football matches, popular television, pantomime, and so on – as, by implication, coarse and lowbrow. I do not plan to argue here for the cultural merits of *I'm a Celebrity, Get Me Out of Here!*, but in insisting that there is no continuity at all between Shakespeare and popular culture, such parodies do perhaps serve to entrench the distinctions between elite and popular cultures rather than providing a site for contestation between them.

This is arguable, however. In Arthur Koestler's formulation of the comic effects of incongruity, jokes are described as 'universes of discourse colliding, frames getting entangled, or contexts getting confused' (1976: 40). For Koestler, comic incongruity implies at once both a separation and a correspondence; he coined the term 'bisociation' to describe the creative act of connecting previously unconnected ideas. Normal social life, he argued, requires us to operate on only one plane of thought at any given time, whereas in order to understand a joke based upon incongruity, its audience must be thinking on two at once. Jokes occur when 'two self-consistent but habitually incompatible frames of reference' collide, allowing 'the perceiving of a situation or idea ... in which the two intersect' (1976: 35). Bisociation between such ordinarily incompatible discourses would, he argued,

> produce a comic effect, provided that the narrative, the semantic pipeline, carries the right kind of emotional tension. When the pipe is punctured, and our expectations are fooled, the now redundant tension gushes out in laughter.
>
> (1976: 51)

For Koestler, laughter becomes a means of escaping our 'automatised routines of thinking and behaving ... the spontaneous flash of insight which shows a familiar situation or event in a new light, and elicits a new response to it' (1976: 45).

Like Koestler, Jonathan Miller perceives laughter as an act of creative thought, seeing great value in the fact that 'it involves the rehearsal of alternative categories and classifications of the world' (1988: 11). In support of this concept of humour, Miller invokes carnival theory and even Bergson (humour, he points out, 'restores us to the more versatile versions of ourselves'; 1988: 16).[10] In this sense, then, the comic incongruities between Shakespeare and the popular forms studied here might in fact be read as joyously creative acts of thought, the disruption of hard-and-fast categories.

Vladimir Nabokov once described parody as 'a game' (1973: 75). Malachi Bogdanov's 2003 Edinburgh Fringe production *Bill Shakespeare's Italian Job* might be best understood in this sense. As its title suggests, the show retold the story of the popular 1969 film *The Italian Job* using (for the most part) lines appropriated from Shakespeare. Thus Gilz Terera, in a Michael Caine accent, would open the show with a set of quotations from *Romeo and Juliet*, before concluding: 'Henceforth, I never will be Romeo – I shall be Charlie Croker.' Later, when asked about the impending heist, he replied with the words of Richard III: 'Plots have I laid, inductions dangerous.' The joy of watching the show, I would suggest, lay in a delighted surprise at the unexpected correspondences between the works of Shakespeare and the film.[11] In press interviews, Bogdanov would unfailingly characterise the production as accessible Shakespeare; he described it as 'the smiling face of Shakespeare' in the *Independent* (28 July 2003), while a *Times* interview quoted him as being 'concerned with trying to develop a younger audience for Shakespeare' (4 August 2003). As with all popular appropriations, however, the show's cultural politics were ambiguous: much like *The Shakespeare Revue*, it certainly rewarded a familiarity with Shakespeare's plays (and, of course, with *The Italian Job*); and since it relied on twisting Shakespearean lines out of shape and out of context in order to fit them to the plot of the film, its success as a 'reappropriation' of Shakespeare must be questionable.

Parody as reappropriation

Shows like *Bill Shakespeare's Italian Job*, then, suggest a synergy between Shakespeare and pop culture at the same time as deriving comic effects from their perceived disparity. A production in which this has been demonstrated with perhaps rather more success is Rick Miller's *MacHomer*. Here, the text of *Macbeth* is delivered by Miller himself in the voices of assorted characters from the animated TV sitcom *The Simpsons* (with Homer, naturally, in the role of *Macbeth*). Miller interweaves the popular personae of *Simpsons* characters with the Shakespearean text, sometimes creating hybrid versions of the characters, at once both Shakespearean and Simpsonian (as the name MacHomer implies), and at others (as in the following extract) implying that the *Simpsons* characters are only temporarily inhabiting the Shakespearean roles:

LADY MACHOMER / MARGE.   All our service
    In every point twice done and then done double.

DUNCAN / MR BURNS.   Malcolm, who's that blue-haired
  mathematician?

MALCOLM / SMITHERS.   Um, Lady MacHomer, sir.

DUNCAN / BURNS.   Ah, Lady MacHomer, eh? Fair and noble hostess,
  We are your guest tonight.
  Conduct me to mine host: we love him highly,
  And shall continue our graces towards him.
  Uh, who wrote this feeble dialogue, anyway?

MALCOLM / SMITHERS.   Um, Shakespeare, sir.

DUNCAN / BURNS.   Shakespeare, eh? Fire that Shakespeare fellow!

MALCOLM / SMITHERS.   Um, he's dead, sir.

DUNCAN / BURNS.   Excellent.[12]

Like Bogdanov's show – and countless other popular appropriations –
Miller's is sold as accessible Shakespeare: the production is advertised
on its official website as 'a terrific trip for middle or high school students sinking their teeth into Shakespeare' (MacHomer 2007), while
reviews quoted on the same site comment on the way in which it
'makes *Macbeth* easier to understand' (*New Jersey Courier-Post*) and 'not
only made the play immediately accessible to a young generation,
it also gave Shakespeare's moral message ironic relevance' (*Ottawa
Sunday Sun*). Unlike Bogdanov's, however, Miller's show relies on a
direct appropriation of a single Shakespearean play (and 85 per cent
of the words in the show are Shakespeare's, according to the website).
In an interview with the *Los Angeles Times*, in fact, Miller suggested
an unexpected correspondence between Shakespeare's title character
and his *Simpsons* replacement: 'It's this guy with this huge ambition
and inability to see beyond the actual act to the consequences, which
is very Homer' (31 March 2001). Dubious as this comparison may
be, the fact that Miller draws it at all is significant. The joke is not
that Homer is a comically inappropriate Macbeth; it is that a cultural
distinction between elite and popular forms renders him a comically
appropriate one.[13]

It is often when pop culture lifts most directly from Shakespeare that
its challenge to his cultural hegemony is most effective. There is, after
all, a wealth of Shakespearean text which has its roots in 'unofficial'
culture, and it has occasionally been reclaimed as such. Two notable
instances were the performances first by the Crazy Gang and then by

the Beatles of the Pyramus and Thisbe scene from *A Midsummer Night's Dream* (5.1).[14] The former were a sketch troupe comprised of popular comedians Chesney Allen, Bud Flanagan, Jimmy Gold, Eddie Gray, Teddy Knox, Charlie Naughton, and Jimmy Nervo, whose shows were a regular fixture in London's West End from 1935 until 1960. Their massively successful revue *These Foolish Kings* ran at the Victoria Palace Theatre from 1956 until 1958, and featured a sketch in which the Gang performed Shakespeare's scene almost verbatim (its only major change was the omission of the interruptions from the aristocratic characters).[15] Performed in the context of a revue, however, the sketch puzzled some commentators. Rather than seeing its inclusion as reappropriation of Shakespeare's scene by popular entertainers, a review in the *Times* suggested the reverse: 'As the Gang have grown in prosperity,' the critic complained, 'they seem to have lost touch a little with the low Cockney humour of the backstreets' (19 December 1956). Similar reactions greeted comedian Frankie Howerd's casting as Bottom in the Old Vic's production of *A Midsummer Night's Dream* the following year: in his autobiography, Howerd suggests that his transition from variety into 'serious' acting served to alienate his fans: 'For me to go from Music Hall to plays, culminating in Shakespeare, gave many people the impression I'd gone all up-market' (1977: 133).

Six years after *These Foolish Kings* closed, the Beatles' performance of the Pyramus and Thisbe scene met with a very different response. Broadcast on the evening of 6 May 1964 as part of the television special *Around the Beatles*, the sketch featured Paul McCartney as Pyramus, John Lennon as Thisbe, Ringo Starr as the Lion, and George Harrison as Moonshine. They performed the scene in full costume (McCartney, for example, in cartoonishly striped doublet and hose) upon an Elizabethan-style thrust stage, surrounded by an audibly excited crowd. The dialogue was broadly Shakespeare's, though it had been heavily cut and was several times ad-libbed upon, and both the Beatles' performance and the audience's response to it exhibited all the festive energy one might associate with the band's live concerts. Their first entrance was met with rapturous applause and cheering, which continued under Trevor Peacock's delivery of the prologue; throughout the scene, McCartney's every sideways look and cheeky grin was greeted with whoops of glee from the crowd. Shakespeare's lines were spoken with the Beatles' own working-class Liverpudlian accents, and their constant interpolations displayed a pronounced irreverence towards the text; Ringo's Lion, for example, departed from Shakespeare with the words:

Then know that I, one Ringo the drummer am; for if I was really a lion, I wouldn't be makin' all the money I am today, would I?

The playful rebellion against cultural authority staged by such ad-libs was met with delighted (and very noisy) cheering from the audience. In his article on the skit, Wes Folkerth draws on Bakhtin to describe the phenomenon as a 'festive uncrowning – not only of Shakespeare himself, but perhaps more importantly, of a certain established way of relating to his works' (2000: 75).

Mention should be made of the sketch's use of staged heckling from certain members of its audience – namely the instrumental band Sounds Incorporated, who made catcalls such as 'Go back to Liverpool!' from an upper balcony throughout the performance. Clearly these put-downs were intended to serve the same dramatic function as the interruptions from the aristocratic characters in Shakespeare's play; interestingly, though, they provoked apparently genuine counter-heckling from members of the audience ('shut up!' is barked back at them throughout). The staged confrontation between the Beatles and their hecklers – and the audience's response to it – encapsulates, I think, something of the cultural forces intent on keeping Shakespeare and pop culture apart, and of pop culture's resistance to those same forces. The taunts suggest not only that the Beatles have no business perform-ing high culture ('Roll over Shakespeare!' was jeered at one point), but also that Shakespeare holds no interest for popular audiences ('What a load of old toffee!' was shouted in a defiantly cockney accent). The crowd's sympathies, however, lay naturally with the Beatles; and when George Harrison's Moonshine stepped out of character to respond to the hecklers, the crowd screamed their approval:

> Look you: all I have to say is to tell you that this lantern is the moon, you see – got it? – I'm the man in the moon; this thornbush here's my thornbush; and this doggy-woggy here's my dog – and if you don't wrap up I'll give you a kick in the – the arse!

This moment is ambiguous for two reasons. First, the audience's sup-port of Harrison against the hecklers indicates an implicit support for the Beatles' appropriation of Shakespeare (a marked contrast from the experiences of Frankie Howerd and the Crazy Gang); clearly, how-ever, they are simultaneously enjoying his obvious departure from the Shakespearean script. Second, while Harrison's ad-lib may have implied a thoroughly transgressive departure from Shakespeare for those unfamiliar with *A Midsummer Night's Dream* – and judging by the audience response, it did – Harrison was in fact merely paraphrasing Starveling's extra-dramatic outburst from the Shakespearean text:

> All that I have to say is to tell you that the lantern is the moon, I the man i'th'moon, this thorn bush my thorn bush, and this dog my dog.

(5.1.252–4)

Harrison's unauthorised interpolation was, ironically, based on an original 'unauthorised interpolation' *authored* by Shakespeare.

The extent to which the audience were aware of Shakespeare's scene, then, determined to a very great extent the cultural meanings of the sketch. Though the Pyramus and Thisbe scene is Shakespeare's own parody of archaic dramatic forms, the Beatles' appropriation of it may have given the impression that they were carnivalising Shakespeare's own work; in this performance, as Folkerth suggests, 'Shakespeare is himself shown to be associated in the popular imagination with the exact kind of bombastic, overblown rhetoric that Bottom employs in this scene' (2000: 77). On the one hand, if the Beatles' audience were at least partially aware of the content and context of the original Pyramus and Thisbe scene, they might have been aware that the performance demonstrated quite clearly a continuity between Shakespeare and pop culture; if, on the other, they were under the impression that the Beatles were sending up Shakespeare's writing, it would have taken on entirely the opposite meaning. The chances are, of course, that the scene was ambiguous enough to sustain both responses, simultaneously and irreconcilably. But ultimately the cultural meanings of the broadcast are virtually impossible to fathom, lying as they do in the perceptions of its audience: and aside from a subjectively judged impression of consensus – which will, in any case, be constantly shifting – there is no way in which one might attempt to pin down such a nebulous phenomenon.

Ambiguous appropriation: Inversion or invocation?

It was perhaps this very ambiguity which made the Beatles' sketch – and other appropriations like it – so very exciting and culturally loaded. As we have noted, one can see something of this ambiguity in most popular appropriations of Shakespeare, and I have quite deliberately postponed discussion of one of the best-known and longest-running contemporary 'Shakespop' parodies until this point in order to explore

Figure 4.1 The Reduced Shakespeare Company performing at the 1987 Renaissance Pleasure Faire in Novato, California. (From left to right, Adam Long, Jess Borgeson, and Daniel Singer.)

it with reference to all the inconsistent and mutually contradictory 'attitudes' of parodic appropriation discussed above.

The Reduced Shakespeare Company's *Complete Works of William Shakespeare (abridged)* began life as a series of comic truncations of *Hamlet* and *Romeo and Juliet* for American Renaissance fairs in the early 1980s (Figure 4.1); these were adapted into a full-length show for the Edinburgh Fringe Festival in 1987, which claimed to present 'all 37 plays by three actors in one hour' (Borgeson, Long & Singer 1994: 119). International touring followed, the show was expanded to two hours, and it ran at London's Arts Theatre from 1992 – and then at the Criterion between 1996 and 2005 – becoming the West End's longest-running comedy. The script was published as *The Compleat Works of Wllm Shkspr (abridged)* in 1994.

On the production's US publicity, an endorsement from *The Today Show* reads: 'If you like Shakespeare, you'll like this show. If you hate Shakespeare, you'll *love* this show!' (Reduced Shakespeare Company 2007). It is a good indicator, I think, of the production's cultural ambiguity. As a popular Shakespearean parody, naturally it encompasses the kinds of confrontation with high culture we encountered earlier: Shakespeare's own writing is mocked, for example when the trio relate the story of a composite Shakespearean comedy ('condensing all sixteen of Shakespeare's comedies into a single play') which is named (among other alternative titles) 'The Comedy of Two Well-Measured Gentlemen Lost in the Merry Wives of Venice on a Midsummer's Twelfth Night in Winter' (Borgeson, Long & Singer 1994: 37). The plot concerns a duke and his family, several pairs of lovers, magic, cross-dressing, and an improbable number of sets of identical twins. Shakespeare's style, too, is lampooned:

> ADAM / JULIET.   Gallop apace, you fiery-footed steeds,
> And bring in cloudy night immediately.
> Come civil night! Come night! Come Romeo,
> Thou day in night! Come, gentle night!
> Come loving, black-brow'd night!
> O night night night night ...
> Come come come come come!
> *(aside to audience)* I didn't write it.
>
> (1994: 20–1)

But as we saw earlier, it is specific practices and cultural uses of Shakespeare which come in for the heaviest mockery. A caricature academic, lamenting falling cultural standards, rapidly descends into fire-and-brimstone evangelism, foreseeing a 'glorious future' in which the *Complete Works* 'will be found in every hotel room in the world!' (1994: 5). Later, in a wickedly accurate spoof of academic jargon, we learn that one of the cast

> has traced the roots of Shakespeare's symbolism in the context of a pre-Nietzschean society through the totality of a jejune circular relationship of form, contrasted with a complete otherness of metaphysical cosmologies, and the ethical mores entrenched in the collective subconscious of an agrarian race.
>
> (1994: 26–7)[16]

Much of the show's humour relies on a perceived incongruity between Shakespeare and popular forms: *Titus Andronicus* becomes a cookery show, *Othello* a rap, and the kings of Shakespeare's histories play a game of American football with the English crown. Throughout the show, a familiarity with Shakespeare is rewarded: the composite-comedy mentioned above, for example, makes oblique reference to every one of Shakespeare's comedies.

For the most part, the Reduced Shakespeare Company play Shakespeare in what has been described above as a 'comically inappropriate' fashion. Lines are garbled and 'misunderstood', and carnivalesque inversions of famous moments abound. 'What's in a name, anyway?' declaims a drag Juliet; 'That which we call a nose / By any other name would still smell' (1994: 16). Again, we see the poetically elevated dragged down to the level of the physical in order to 'demystify': an irreverent chorus suggests a purely sexual motivation behind Romeo's actions ('in a scene of timeless romance, / He'll try to get into Juliet's pants'; 1994: 14), and later in the same sequence, Juliet attempts to perform the balcony scene while standing on another cast member's shoulders (culminating in the inevitable comic pratfall).

There is, however, a very strong sense that despite their mockery – or through it, perhaps – the Reduced Shakespeare Company are attempting to snatch Shakespeare back from the highbrow and the academic, and to reclaim his work for popular culture. Their adaptation of *Romeo and Juliet* remains surprisingly 'straight' over the 50 lines or so of its first scene, and aside from some deliberately crude paraphrasing during the servants' brawl, draws exclusively on text from Shakespeare's 1.1. And though their radically truncated *Hamlet* is a riot of anarchic energy, it features a few moments in which the play is presented as a source of truly affecting dramatic power. While discussing what can safely be cut from the play, for example, Daniel Singer remembers the 'What a piece of work is man' speech from 2.2. Adam Long proceeds to recite lines 297–310 of this speech *simply, quietly and without a trace of interpretation. You can hear a pin drop*' (1994: 81). There is, of course, a punchline: 'So,' says Singer after a pause, 'we'll skip that speech and go right to the killing'. But the joke is predicated on the suggestion that Long's delivery has captured something profound about the speech which makes a desire to skip to 'the killing' seem comically shallow, and it gestures towards a popular appropriation of Shakespeare which retains the plays' intellectual force and emotional power.

Parody as criticism

As we noted at the beginning of this chapter, the distinction between parody and other forms of metafiction might be said to lie in its inflection, or attitude. Since the text contains no inherent 'attitude', though, this inflection can be determined only in the audience response, and as we saw with the Beatles, this is impossible to pin down. Ultimately, then, the boundary separating parody from other forms of textual recycling and allusion is a very blurred one. Linda Hutcheon extends her definition of parody to encompass non-comic forms of intertextual reference: parody, by her argument,

> is a form of imitation, but imitation characterised by ironic inversion, not always at the expense of the parodied text. ... Parody is, in another formulation, repetition with critical distance, which marks difference rather than similarity.
>
> (1985: 6)

In Hutcheon's definition of parody as 'repetition with critical distance' we find very strong echoes of the theorist whose ideas concerning popular theatre have underpinned every chapter of this book so far; Hutcheon herself notes this similarity, in fact. 'Like Brecht's *Verfremdungseffekt*,' she remarks, 'parody works to distance and, at the same time, to involve the reader in a participatory hermeneutic activity' (1985: 92). Here, parody has the potential to become a form of Brechtian alienation, its ironic distancing allowing its audience to reassess their attitude towards its source text.

Brecht adapted Shakespearean texts several times during his career, and though none of these are usually classed as 'parodies', they are certainly parodic in Hutcheon's broader sense of the term. His appropriations included radio adaptations of *Macbeth* (1927) and of *Hamlet* (1931), now sadly lost; *Die Rundköpfe und die Spitzköpfe* ('The Roundheads and the Peakheads'), a very loose adaptation of *Measure for Measure* (1936); a set of 'rehearsal scenes' based around *Macbeth, Hamlet,* and *Romeo and Juliet* (1939); some appropriation of Shakespearean dialogue in *Der aufhaltsame Aufstieg des Arturu Ui* (*The Resistible Rise of Arturo Ui*, 1941); and, finally, his adaptation of *Coriolanus*, which remained unfinished at his death in 1956. In almost all these cases, Brecht aimed at an interrogation of the hegemonic politics he saw embodied in Shakespeare's plays.

Brecht's attitude towards Shakespeare was deeply ambivalent. He admired a great deal of Shakespeare's dramaturgy, describing his theatre as one 'full of A-effects' (Brecht 1965: 58). Shakespeare's great failing in Brecht's eyes, however, was that he portrayed human suffering as both ennobling and inevitable, and that he failed to address its social and political causes. In his *Short Organum for the Theatre*, Brecht explains that

> [t]he theatre as we know it shows the structure of society (represented on the stage) as incapable of being influenced by society (in the auditorium). ... Shakespeare's great solitary figures, bearing on their breast the star of their fate, carry through with irresistible force

their futile and deadly outbursts; they prepare their own downfall; life, not death, becomes obscene as they collapse; the catastrophe is beyond criticism.

(1977: 189)

According to Brecht, then, it was the role of the theatre to debunk such notions. As 'the Philosopher', Brecht's spokesperson in *The Messingkauf Dialogues*, puts it:

> THE PHILOSOPHER.   The causes of a lot of tragedies lie outside the power of those who suffer them, so it seems.
>
> THE DRAMATURG.   So it seems?
>
> THE PHILOSOPHER.   Of course it only seems. Nothing human can possibly lie outside the powers of humanity, and such tragedies have human causes.
>
> (Brecht 1965: 32)

The implication for the performance of Shakespeare's plays, of course, is that some sort of intervention becomes necessary. Thus, on the matter of performing *King Lear*, the Philosopher explains:

> If you're going to perform it on the new principle so that the audience doesn't feel completely identified with this king, then you can stage very nearly the whole play, with minor additions to encourage the audience to keep their heads.
>
> (1965: 62)

For Brecht, Shakespeare's other major fault was his bias towards the ruling classes: the suffering of the nobility is usually emphasised over that of the workers. Brecht's sympathies were always with the proletarian characters. 'What you cannot have,' the Philosopher continues,

> is the audience, including those who happen to be servants themselves, taking Lear's side to such an extent that they applaud when a servant gets beaten for carrying out his mistress's orders as happens in Act 1 scene 4.
>
> (1965: 62)

Brecht's impulse to emphasise the social imbalance in Shakespeare's plays becomes the overriding political feature of his Shakespearean appropriations. His 'rehearsal scenes' for *Romeo and Juliet* – written in 1939 for his wife Helene Weigel's acting classes – provide perhaps the most explicit examples. These scenes, intended for rehearsal just before the balcony scene of 2.1, depict the play's title characters exploiting their social superiority for entirely selfish reasons. Romeo, anxious to palm off his former mistress Rosalind with a parting gift before turning her 'out into the streets', is seen trying to extort money from an impoverished tenant; Juliet, meanwhile, is shown demanding that her Nurse miss an assignation with her lover Thurio, in order to stay in and cover for Juliet during her own tryst with Romeo. In both cases, Romeo and Juliet's 'feelings' are shown to be rooted in the privileges of the upper class:

ROMEO.   what do I know about estates – I'm burning up!

TENANT.   And we are hungry – sir.

ROMEO.   Stupid! Isn't there any way to reason with you? Don't you animals have any feelings? Then off with you, the sooner the better.

(1967: 108)

JULIET.   Then you certainly don't love him.

NURSE.   What do you mean? Don't you call it love when I want so much to be with him?

JULIET.   But it is an earthly love.

NURSE.   But that's beautiful, earthly love, don't you think?

JULIET.   Of course. But I love my Romeo more than that, I can tell you.

(1967: 109)

Both the Tenant and the Nurse remain on stage in the background during the balcony scene, as a reminder of the social inequalities which provide the context for Romeo and Juliet's otherwise seemingly classless romance.

In his adaptation of *Coriolanus*, meanwhile, Brecht follows Shakespeare very closely, but his careful rephrasing of the dialogue indicates an important shift in perspective. Taking issue with the way in which Shakespeare's plebeians 'are shown as comic and pathetic types (rather than humorous and pathetically treated ones)' (1977: 255), he deems it essential to make Menenius Agrippa's political manipulation of them explicit:

> We've got to show Agrippa's (vain) attempt to use ideology, in a purely demagogic way, in order to bring about that union between plebeians and patricians which in reality is effected a little – not very much – later by the outbreak of war. ... Agrippa's ideology is based on force, on armed force, wielded by Romans.
>
> (1977: 258)

Thus, for example, what reads in Shakespeare as follows:

> Nay, these are all most thoroughly persuaded,
> For though abundantly they lack discretion,
> Yet are they passing cowardly.
>
> (1.1.199–201)

– is rendered in Brecht's version like this:

> Let be. I've won these fellows over, stopped them
> With a fairy tale. Though to be sure it was not
> The sword of my voice but rather the voice of your sword
> That toppled them.
>
> (Brecht 1973: 64)[17]

Brecht arguably had greater success as a Shakespearean critic than as an author of Shakespearean appropriations in his own right: incisive as his rehearsal scenes are, they were never intended for inclusion in a public performance, and his *Coriolanus* does not stage a commentary on Shakespeare's play so much as 'amend' it in order to provide a commentary on the actions of the *characters* (Brecht discusses 'amending' Shakespeare in his 'Study of Shakespeare's *Coriolanus*'; 1977: 259). Shakespeare's perceived faults, as the example above shows, disappear in the rewrite.

However, Brecht's ideas paved the way for countless other practitioners' political adaptations of Shakespeare, among them Edward Bond's *Lear*, Arnold Wesker's *The Merchant* (later retitled *Shylock*), Augusto Boal's *A Tempestade*, Welfare State's *King Real*, and Charles Marowitz's various Shakespearean 'collages'. In each case, adaptation becomes a means of critical commentary upon the original; as Marowitz explains of his *Hamlet*, 'built into the exercise was not only the play itself, but the adaptor's attitude to the play' (1991: 19). In some cases, theatrical commentary has drawn quite strongly from the tactics of literary criticism. In its juxtaposition of passages of *The Merchant of Venice* with other historical texts (including passages from the Bible, *The Jew of Malta*, and Edward I's 1290 decree expelling Jews from England), Gareth Armstrong's recent one-man play *Shylock* (1999) was making use of a favourite strategy of new historicist criticism.

Critical commentary in popular appropriations

The examples cited in the section above are all, by Hutcheon's definition, broadly 'parodic' – in that they allude to Shakespeare's plays with ironic distance – and 'popular' in the sense that politically, they are taking the part 'of the people'. To present a progressive and sympathetic account of plebeian characters, however, is not necessarily the same thing as to produce a piece of popular theatre. As Marowitz acknowledges, 'one of the prerequisites for Shakespearian collage is the audience's general familiarity with the play' (1969: 15): a sophisticated intertextual knowledge is required on the part of the audience. Thus, in the examples above, Shakespeare becomes the site of a political confrontation between hegemonic and popular forces; but culturally, arguably, the landscape upon which the battle is waged remains that of the elite.

We will turn, finally, to a selection of full-length theatrical appropriations of Shakespeare which were not only parodic but which also aligned themselves quite firmly with popular culture. These appropriations signify a disjunction not so much from the Shakespearean *text* (though depart from it they certainly do) as from 'Shakespeare', the established cultural force. We saw in Chapter 2 that Shakespearean productions are often insistent, as Alan Sinfield put it, 'that what they are presenting is *really Shakespeare*'; Sinfield describes this phenomenon as 'the importance of being Shakespeare' (2000: 185). Naturally this throws up a number of questions as to where 'Shakespeare-ness' lies. A recent 'wordless' production of *Macbeth* by American company Synetic Theater, for example, provoked a lively debate within the Shakespearean academic community as to whether or not such an undertaking could possibly constitute 'Shakespeare'; some made claims that the show 'profoundly explores Shakespearean art', while others objected that a wordless production could never be 'an act of aesthetic fidelity' (Galbi, Manger, Drakakis et al. 2007). Arguments in favour of the production's status as 'Shakespeare' tended to imply that the 'essence' of the play lay not only in Shakespeare's words, but also in the more theatrical aspects of his dramaturgy – his structuring, his stagecraft, his use of archetypes – even simply his plot (a particularly problematic attribution, since Shakespeare drew much of this from Holinshed).[18] John Drakakis joined the debate to debunk such claims for Shakespeare's transcendence of language, however, arguing that the Shakespearean text *was* in fact present (albeit invisibly):

> Presumably a 'wordless' Shakespeare can only work when you recognise what is missing. In other words, it is a defamiliarising gesture that ranks alongside Stoppard's *Dogg's Hamlet* or the work of the Reduced Shakespeare Company, or that of Charles Marowitz. It is now up to those who champion wordless Shakespeare to tell us what it is that this contributes to our understanding of the play.
>
> (Galbi, Manger, Drakakis et al. 2007)

Drakakis' analysis, of course, corresponds very strongly with the ideas laid out in this chapter regarding Shakespearean appropriation as a form of metatextual commentary. I would argue, though – not having seen the production – that as a piece of physical theatre which was by all accounts a visually captivating synthesis of drama, movement and music, Synetic Theater's *Macbeth* may well have had value outside of its contribution 'to our understanding of the play'.[19]

The argument as to whether or not the production was 'faithful' to Shakespeare was, I would suggest, an ultimately meaningless one. 'Fidelity' is not a particularly useful concept in discussion of Shakespearean appropriations, where the intention is generally to signal a departure from (instead of, or as well as, an affinity with) the Shakespearean text. There is, in any case, no such thing as a Shakespearean 'essence', and when such a concept is invoked, what is being brought into play is in fact whichever set of values the speaker wishes to imbue with Shakespeare's cultural weight. Buying into 'Shakespeare' in this sense frequently means, as Lanier suggests, buying into the 'principles of aesthetic and moral cultivation for which Shakespeare is symbol and vehicle': such a 'Shakespeare', he argues, 'springs from an affiliation between Victorian moralisation and high modernist aesthetic elitism' (2002: 100).

The adaptations which follow remain ambivalent about their claims to 'Shakespearean' status, and might be seen to act as ironic commentaries, not so much on the Shakespearean texts upon which they are based, but on the cultural uses to which those texts – and the name 'Shakespeare' – are put. They might be considered examples of carnivalesque 'uncrownings' of Shakespeare, and (to use Drakakis's term again) 'demystifications' of his cultural authority. At the same time, however, they might be seen as pointing towards an 'authentically' popular Shakespeare (no less constructed, of course, than the elitist Shakespeare), and as engaged in an attempt to *redefine* Shakespeare's cultural meanings. If parody marks 'difference rather than similarity' (Hutcheon 1985: 6), then these incorporate both parodic and non-parodic elements.

Contrasting *Cymbelines*

'Naughty Kneehigh,' began John Peter's review of Kneehigh Theatre's *Cymbeline* (one of the more provocative inclusions in the RSC's 'Complete Works' season): 'their show is not, as the programme says, by Shakespeare. It's a loving send-up' (*Sunday Times*, 1 October 2006).[20] Indeed, the production was a very loose adaptation: there were scattered quotations from Shakespeare's original in each scene, but these accounted for a relatively small proportion of the whole script. Structurally, however, it followed Shakespeare's play almost scene-for-scene, often in line-for-line paraphrases. Its ambiguous and dialogic attitude towards its source renders it arguable, I think, that Kneehigh's show was significantly more than merely a 'send-up'.

Most of the text was written in a contemporary, 'unofficial' idiom – often quite defiantly so. 'Who the fuck're you?' said Cloten to Iachimo, in a particularly non-Shakespearean exchange (Rice & Grose 2007: 32); later, attention was drawn to the play's disjunction from traditional Shakespearean performance modes when Posthumus informed Iachimo, 'You can't say "cock" – we're in Stratford-upon-the-Avon!' (ad-lib). Elsewhere, Shakespearean quotations were deliberately garbled: misreading the sentence in Posthumus' letter, 'Thy mistress hath played the strumpet in my bed' (a very slightly altered line from the original at 3.4.21–2) as an accusation that she had 'played the *trumpet* in my bed', Hayley Carmichael's Imogen tearfully objected, 'I can't even *play* the trumpet!' (2007: 43). Local references were added when the production toured; a note in the script indicates a reference to '*whatever place-name is deemed appropriate to venue*' (2007: 41). Perhaps the most mischievous line was spoken by the production's non-Shakespearean chorus figure, a Cornish widow named 'Joan Puttock' (her name presumably a reference to Imogen's use of the word 'puttock' in the original at 1.1.141).

Figure 4.2 Mike Shepherd as 'Joan Puttock' in Kneehigh's *Cymbeline* (2007).

Having related a long list of relevant plot details to the audience, the character remarked: 'It's like a Shakespeare play!' (2007: 14) (Figure 4.2).

Despite these deliberate disjunctions from the official discourses of the original play, however, many of Shakespeare's lines slotted so seamlessly into Emma Rice and Carl Grose's script that it might have been difficult for audience members to tell them apart. Pisanio and Imogen's exchange –

PISANIO.   Since I received command to do this business
    I have not slept one wink!

IMOGEN.   Oh well, do it, and to bed, then!

    (2007: 43; very slightly altered from 3.4.99–100)

– coming as it did after a breathless, pantomime-style chase through the audience, barely seemed a switch in register at all, and received one of the biggest laughs of the night. Naturally much of the density of Shakespeare's text was lost, but many of his most striking lines were kept intact, and in cutting virtually all of the play's more impenetrable language, Kneehigh 'retold' its complicated story with remarkable clarity. Shakespeare's imagery was augmented for a modern audience: the famous passage, 'Golden lads and girls all must, / As chimney-sweepers, come to dust' (4.2.263–4), for example, was subjected to a minor but significant change, replacing its relatively archaic metaphor with one to which contemporary audiences might have more access; thus 'chimney-sweepers' became 'dandelions'. But John Pitcher's note in the Penguin edition of *Cymbeline* makes it clear that this is less of a departure from Shakespeare's text than it might first appear: 'chimney-sweepers' was in fact a term used in Warwickshire dialect do describe dandelions when they went to seed (4.2.262–3n.). As Valerie Wayne, the Arden3 editor of the play, notes in her article on the production for *Shakespeare Quarterly*: 'Kneehigh does its research when altering Shakespeare's text' (2007: 236). The script itself gives a strong sense that Rice and Grose carefully assessed each line of Shakespeare's play and included it in their adaptation when, and only when, they found it more theatrically effective than any possible replacement.

This chapter opened with quotations from two 2007 adaptations of *Cymbeline* – one from Kneehigh's, and the other from Cheek by Jowl's. The latter were good enough to send me a copy of their unpublished script, and it shows them to have undergone a surprisingly similar process. Shakespeare's more impenetrable language is jettisoned here, too: Cheek by Jowl's script is just over half the length of Shakespeare's, and much of what is left is rewritten (generally for the sake of clarity). Thus, for example, Iachimo's reference to

> tomboys hired with that self exhibition
> Which your own coffers yield; ... diseased ventures
> That play with all infirmities for gold
> Which rottenness can lend to nature
>
> (1.6.123–6)

becomes simply 'strumpets hired with diseased gold' (Cheek by Jowl 2007: 23). Lines are added to the beginnings of scenes to clarify context: 1.4 now begins 'Know you aught of the young Briton who is newly come from Rome?' (2007: 11), and Posthumus makes clear the significance of his 'bloody handkerchief' with the lines 'So Imogen is dead ... / Pisanio hath sent this bloody sign of it', before beginning Shakespeare's 5.1 (2007: 79). The extract quoted at the beginning of this chapter adds dialogue to what, in Shakespeare, is a dumb show (*'enter again in skirmish Iachimo and Posthumus: he vanquisheth and disarmeth Iachimo, and then leaves him'*; 5.2).

The review by John Peter quoted above commented that in Kneehigh's production, the forest scene between Imogen and Pisanio was 'as touching as in any straight production' (*Sunday Times*, 1 October 2006). Taking my cue from this observation, I set about comparing this scene in Kneehigh's 'send-up' with its equivalent in Cheek by Jowl's uncontestedly 'straight' production. The following extract is Kneehigh's truncation of Shakespeare's 3.4.40–96:

> IMOGEN.   False to his bed? What does he mean by false?
> To lie in wait there and to think on him?
> And cry myself awake? That's false is it?
> Me? False? Iachimo said he had found his
> Way between the thighs of Italian whores ...
> Now I'm thinking he was speaking the truth!
> Come, then. Finish the job he asked of you.
> Do your master's bidding, Pisanio!
> Hit the innocent mansion of my heart!
> Fear not, 'tis empty of all things but grief!
>
> PISANIO.   I can't!
>
> IMOGEN.   Come on! Kill me!
> The lamb entreats the butcher!

> (2007: 43)

Kneehigh cut around 368 words from the Shakespearean text, and though most of the remaining lines were slightly rephrased ('What is it to be false?', for example, became 'What does he mean by false?') only six lines were significantly altered (most notably those between 'Iachimo said ...' and 'Finish the job he asked of you.'). Cheek by Jowl, meanwhile, rendered the passage as follows:

> IMOGEN.   False to my husband's bed? What is it to be false?
> To lie in watch there and to think on him?
> To weep 'twixt clock and clock? If sleep charge nature,
> To break it with a fearful dream of him
> And cry myself awake? That's false to's bed, is it?
>
> PISANIO.   Alas, good lady!
>
> IMOGEN.   I false! Some jay of Italy
> Whose mother was her painting, hath betray'd him:
> Poor I am stale, a garment out of fashion;
> And, for I am richer than to hang by the walls,
> I must be ripp'd: – to pieces with me! – O,
> Men's vows are women's traitors!

PISANIO.   Good madam, hear me.

IMOGEN.   Thou, Posthumus dost belie all honest men;
Goodly and gallant shall be false and perjured
From thy great fall. Come, fellow, be thou honest:
Do thy master's bidding: when thou see'st him,
A little witness my obedience: look!
I draw the sword myself: take it, and hit

PISANIO.   Alas. Good lady!

IMOGEN.   Why, I must die;
Come here's my heart.
The lamb entreats the butcher.

(Cheek by Jowl 2007: 51–2)

This extract cuts about 287 words, and though most of the lines which remain in their script are exactly as they appear in Shakespeare's text, two are significantly reworded ('thou, Posthumus, / Wilt lay the leaven on all proper men' is rephrased as 'Thou, Posthumus dost belie all honest men', and in both extracts, Pisanio's 'Hence, vile instrument, / Thou shalt not damn my hand!' is replaced with a completely different line). Both productions, then, altered the scene in a similar fashion: both cut more of Shakespeare's words than they left in, and both inserted paraphrases of their own. In neither, essentially, was Shakespeare's text left in anything like its integrity. The significant difference, of course, is that Cheek by Jowl's departures from Shakespeare were clearly intended to be unnoticeable as such, while Kneehigh's made its changes self-evident. The fact that Cheek by Jowl's production made precisely the same sorts of changes to the scene as Kneehigh's indicates perhaps that 'straightness' lies not so much in any objective standard of textual fidelity, as in that unquantifiable variable, 'attitude'. Cheek by Jowl chose a predominantly 'textual' attitude; Kneehigh, clearly, a 'metatextual' one.

The extent to which Kneehigh's production could stake its claim to status as 'straight' Shakespeare was widely debated in the newspaper reviews. On the one hand, claims for an element of fidelity were made by critics such as the *Independent*'s Paul Taylor, who pointed out that as the 'most meta-theatrical' of Shakespeare's late romances, *Cymbeline* was 'right up the street of Kneehigh who revel in the blackly comic clash of tones and anarchic knowingness' (though he did note that 'Kneehigh's brand of jokiness suits some of the strands in this odd tapestry better than others'; *Independent*, 27 September 2006). Michael Billington, on the other hand, while praising the production's 'ingenious visuals', complained that it taught him 'nothing new about the play' and that it 'ducks the challenge of making Shakespeare live through his language'. 'I felt we were being asked to celebrate Kneehigh's cleverness,' he concluded, 'rather than explore Shakespeare's own mysterious, experimental genius' (*Guardian*, 23 September 2006).

Academic reactions to Kneehigh's production were, on the whole, more sensitive to its ambiguity. Valerie Wayne took issue with Billington's conclusion, arguing in *Shakespeare Quarterly* that 'the production does not require that we make that choice, for it draws on the excellence of both.' She found herself 'impressed by how fully Kneehigh conveyed not the letter of the text, but its spirit':

> Rather than simplifying the characters or plot to elicit easy enthusiasm from an audience, the company carefully reframed Cymbeline within a contemporary idiom yet generated responses that are remarkably close to those elicited by the very best productions of Shakespeare's play.
>
> (2007: 231–2)

Whereas Wayne constructed the show in both textual and metatextual terms, ('Kneehigh's adaptation is and is not Shakespeare,' she argued; 2007: 230), Michael Dobson's account in *Shakespeare Survey* focused more fully on the production's metatextual commentary. Noting its tendency to quote Shakespeare's play directly only before immediately undercutting or parodying it, Dobson compared the production with 'the best Victorian burlesques', in that 'it has some canny things to say about the play on which it draws' (2007: 311). Interestingly, unlike many of the critics in the press who lamented the production as, for example, a 'dumbing down of the Bard's work' (*The Stage*, 27 September 2006), both Wayne and Dobson drew attention to the production's potentially elitist appeal to Shakespeare initiates only. Dobson expressed concerns that Kneehigh's show would not 'be nearly as amusing to an audience less familiar with their Shakespeare than are the usual crowd at the Swan' (2007: 310), while Wayne worried that viewers who did not know the play 'might not have appreciated how cleverly and carefully it updated the original' (2007: 231).

What is noticeable in many of the production's reviews, whether sympathetic or hostile, is that they position the text as the absolute and ideal original by which Kneehigh's production must be judged; the central question one must address in assessing the show's value, it would seem, is whether or not it had in some sense been true to Shakespeare's play. But clearly Kneehigh were not attempting to participate in the dubious exercise of creating a definitive theatrical embodiment of Shakespeare's text. In a sense, much of the production's drama resided in the confrontation it staged between the discourses of official 'Shakespeare' and the company's own unofficial metanarrative; liberties were being taken, and the now-traditional rituals of watching and performing Shakespeare were being undermined. An element of the thrill of unsanctioned performance lay in the production's challenge to the institutionalisation of Shakespeare, a challenge which felt particularly 'tendentious' given its geographical situation at the RSC's Swan Theatre, in the heart of Shakespeareland. It was deliberately flouting the implied rules of the Complete Works season, blurring the advertised, RSC-imposed distinction between 'The Plays' (supposedly 'straight' productions, though these included several foreign-language versions and a musical adaptation) and 'The Responses' (mostly new plays, though they included an experimental version of *Twelfth Night* and a *Hamlet* performed by puppeteered 'Tiny Ninjas'). Kneehigh's production was clearly designed to be contentious, and indeed it was: when I saw it, I overheard several audience members muttering angrily during the interval ('It's too far from the original Shakespeare', said one), and there were several conspicuously empty seats in the previously sold-out auditorium for the second half. A member of the Front of House staff told me afterwards that one outraged audience member had stormed out into the foyer during the show and physically threatened both the House Manager and actor Mike Shepherd. I would suggest that perhaps what upset the traditionalists so very much was precisely this ambiguous cultural status: the production was neither 'straight' play nor parody, neither unambiguously 'Shakespeare', nor a mere 'response'.

Remixing *The Comedy of Errors*

We will turn now to two final examples of such blurring of cultural categorisations. Both were popular appropriations of *The Comedy of Errors*, and both made use of an interplay between the sorts of official and unofficial registers examined in Chapter 3. The first example is the Flying Karamazov Brothers' production, which met with widespread (though by no means universal) acclaim in the US in the 1980s. Originally commissioned by Gregory Mosher for Chicago's Goodman Theater in 1982, it was revived in New York at the Lincoln Center Theater when Mosher was artistic director there in 1987.[21] Four members of the popular five-man juggling troupe the Flying Karamazov Brothers played the central roles: Howard Jay Patterson ('Ivan') and Paul Magid ('Dmitri') were the two Antipholuses, Randy Nelson ('Alyosha') and Sam Williams ('Smerdyakov') the Dromios. The production featured a supporting cast composed mainly of jugglers, clowns, musicians, and acrobats.

It began with the kinds of challenges to the authority of the Shakespearean text we have seen throughout this chapter, opening, somewhat daringly, with a six-minute long sequence of silent clowning and audience interaction by the mime Avner Eisenberg (stage name 'Avner the Eccentric'). Following this, a second performer entered, to deliver an almost entirely non-Shakespearean paraphrase of 1.1's exposition:

> Welcome to Ephesus! Unless of course you're from Syracuse. We don't like them very much. We caught a man from Syracuse today. He walked into our square. Foolish fellow! We'll hang him high! Hang till he dies, then take his goods, too. How could he dare? He just walked into our square! He could pay off the Duke, but he hasn't got a ducat. Pay the Duke by five o'clock, or kiss his ass goodbye and fuck off to the gallows. He claims he's looking for his sons. Twin sons he lost in tempest tossed! Heh, he lost his two sons, what a clown! *(seeing the clown's red nose)* Oh, no offence.

The above was accompanied by visual illustration – Egeon entered with a lasso around his neck and a placard around his torso proclaiming 'SYRACUSE MERCHANT NEEDS BAIL', while the two Antipholuses appeared from behind revolving doors, twirling their moustaches identically. After some comic business involving the revolving doors, the passage concluded with the warning:

> In Syracuse, you dress in a tie;
> In Ephesus, you juggle – or die!

There followed a sequence of music, juggling, and dancing, culminating in the beginning of Shakespeare's 1.2; over nine minutes of the production had elapsed before a single word of Shakespeare's script was spoken. From then on, however, the tables were turned, and aside from a few cuts and interpolations, Shakespeare's script was delivered pretty much as written.

In many senses, then, this was close to what might be considered 'straight' Shakespeare. However, the Shakespearean text was almost constantly undercut by the physical activities with which it was accompanied. Here, Antipholus of Syracuse's description of Ephesus as 'full of cozenage, / As nimble jugglers that deceive the eye' (1.2.97–8) was taken at face value: thus every character from Ephesus was continuously performing a circus act. The dialogue in 2.1, for example, was delivered while tap-dancing; Adriana's diatribe at Antipholus (2.2.113–49) became a knife-throwing act. Short acts of trapeze, juggling, clowning, and so on were interposed between each scene, and Shakespeare's dialogue was frequently interrupted by the audience's applause at juggling sequences. Shakespeare's text was quite literally 'carnivalised'.

The production's carnivalisation of Shakespeare as a cultural icon was also manifest, since the immediately recognisable figure of Shakespeare himself was incorporated throughout the performance, played as a silent clown by the fifth Karamazov, Timothy Daniel Furst ('Fyodor'). When Avner the Eccentric swept the detritus from his opening act into an onstage trapdoor, Shakespeare emerged from it immediately afterwards, angrily clutching a copy of the script. Later, he would appear during one of the inter-scene circus acts, balancing an electric guitar from his chin. In 2.2, Antipholus and Dromio of Syracuse were forced to demonstrate their circus credentials in order to blend in with the natives of Ephesus. They read lines 53–63 (Shakespeare's jokes concerning 'basting') from an oversized joke book; their onstage audience of jugglers, however, were unamused by their 'act', and surrounded them menacingly. After lines 74–7 (the now woefully unfunny sequence of jokes about baldness), the onstage audience were still unimpressed; the figure of Shakespeare, however, could be seen looking down on the scene from an upper balcony and laughing hysterically. Like other productions discussed throughout this book, the Flying Karamazov Brothers' show was displaying a deliberate, ironic, and playful disjunction from 'Shakespeare', and, by implication, from his cultural associations.[22]

The show had a strong thread, too, of intertextual reference. Occasional interpolations and topical references were added to the Shakespearean text, but most striking were its visual and musical quotations from pop culture. Arthur Holmberg's review for the *Performing Arts Journal* describes this aspect of the production with some outrage:

> Rags and tatters of pop art performance styles dissociated from their original context and totally alien to Shakespeare's text flashed past our glazed eyes in manic profusion: *Evita, Jaws, West Side Story, Gone With the Wind*, Duke Ellington, Johnny Rotten, Howdy Doody, Post-New Wave Punk, the Muppets, Mayor Jan Byrne.
>
> (1983: 55)

Holmberg's concern, of course – like many of the critical objections to the popular appropriations of Shakespeare studied here – was that these additions did not stem in any way from the Shakespearean source. His review concluded with a plea for 'directors and actors of vision and understanding who can transcend the current theatrical schizophrenia that simultaneously reveres and disdains the past' (1983: 55). What Holmberg evidently saw as (or at least, hoped would be) a passing theatrical fad, has, however, become increasingly accepted within not only contemporary theatre practice but also culture more generally. The Flying Karamazov Brothers' production perhaps marks the beginning of a form of Shakespearean appropriation which is defiantly disregarding of cultural categories in its intertextuality, and one which coincides with a more general cultural shift.[23]

Where in the 1950s the Crazy Gang or Frankie Howerd encountered cultural resistance from both sides in their attempted fusions of elite and popular forms, such firm boundaries no longer exist. Dentith describes modern popular culture as a 'karaoke culture', characterised by an 'endless and voracious circulation of cultural material'. 'In a world without cultural hierarchies,' he explains,

> parody here certainly is not – or not only – parody of the 'high' by the 'low', for it more typically fixes upon other products of popular culture itself, as one comedian parodies another, as pop musicians and disc jockeys sample and remix each other, as indeed karaoke itself offers the chance to mimic or act out the incessantly reproduced voices of popular music.
>
> (2000: 184)

Today, perhaps, Shakespeare has become simply another cultural source from which intertextual reference can be picked: thus one found an ironic visual quotation of Tarantino's *Reservoir Dogs* (1992) in Creation Theatre's *Much Ado About Nothing* (2004), just as one finds *Hamlet* references in the Schwarzenegger vehicle *Last Action Hero* (1993). The most persuasive proof of this shift might be the existence of an entire subgenre of American teen films: the high-school Shakespeare adaptation. Taking their lead, arguably, from the twin sources of Baz Luhrmann's phenomenally popular *Romeo + Juliet* (1996) and *Clueless* (1995), Amy Heckerling's trendily ironic updating of Jane Austen's *Emma*, films such as *10 Things I Hate About You* (*The Taming of the Shrew*, 1999), *O* (*Othello*, 2001), *Get Over It* (*A Midsummer Night's Dream*, 2001), and *She's the Man* (*Twelfth Night*, 2006), relocate the plots of their Shakespearean sources into contemporary American high school settings. The adaptations are extremely loose ones, of course, but generally make frequent reference to their Shakespearean originals throughout.

Our final example makes similar use of the crumbling of cultural categories which has taken place over the last half-century. The Off-Broadway and subsequent Edinburgh Fringe hit *The Bomb-itty of Errors*, developed from a New York University thesis project, was a hip-hop version (advertised as an 'ad-RAP-tation') of Shakespeare's play. Like many of the productions discussed above, it consisted of Shakespeare's lines rephrased into a new idiom; thus, for example, the exchange –

ADRIANA.   How if your husband start some other where?

LUCIANA.   Till he come home again, I would forbear.

(2.1.30–1)

– was reinvented along these lines:

ADRIANA.   What would you do if your husband came back late?

LUCIANA.   I'd stand at the door butt-naked and wait.

The production also used elements of improvisation and audience interaction. Like Kneehigh's *Cymbeline*, it also featured much direct incorporation of Shakespeare's words, and Georgina Brown noted in her review that she felt it 'at once sends up and celebrates Shakespeare's style' (*Mail on Sunday*, 11 May 2003). Shakespearean quotations, Lanier explains in his account of the show, were 'woven into a mix of other cultural samplings, contemporary songs, timely allusions, hip-hop and media clichés, and the like' (2002: 78). The implication, argues Lanier, was not of a comic discrepancy between the forms of Shakespeare and rap, but rather of a synergy between them:

[B]oth are poetry designed for performance, not the page; both feature language delivered against a strong metrical beat and display a mastery of rhythmic effects; both use what is for mainstream speakers of English a largely non-standard vocabulary, dense in allusions; both are self-consciously virtuosic in their wordplay.

(2002: 74–5)

This analysis does perhaps need some qualification. While it may be flouting the culturally constructed divisions between 'high' and 'low' art forms inherited from previous generations, I am not certain that a production like this can suggest those divisions do not exist: the very act of disregarding them is an act of defiance and of cultural contestation.

Like many other productions staging a radical, witty, deconstructive appropriation of Shakespeare, *The Bomb-itty of Errors* was described in one review as 'Shakespeare for the MTV generation' (*Spectator*, 10 May 2003).[24] This description frequently seems designed to appear somewhat condescending (if not downright disparaging), the implication being that this demographic is incapable of deep engagement and can manage only bite-sized, highly simplified cultural chunks. But I would suggest there is something more in the analogy, intended or not. We have arrived at a cultural moment where popular forms' means of engagement with 'high art' have extended beyond straightforward quotation or parody. Audiences of the 'MTV generation' may indeed be easy to bore and quick to judge, fingers ever-poised to hop to the next channel, but by the same token they are fluent in their familiarity with diverse and often abstruse cultural references, and expert in their ability to switch from one mode of reception to another. In the emotional responses of this audience, sincerity and insincerity jostle for pride of place. While such an audience is nothing new in popular culture, its effect on mainstream Shakespearean performance – and, by implication, on Shakespearean theatre scholarship – has been increasingly evident over recent years. The potential for a dynamic engagement between this audience and Shakespearean performance remains largely unrealised, but as *Bomb-itty*, *MacHomer*, the Reduced Shakespeare Company and their ilk show, the beginnings are there.

A question underlying many of the critical responses cited in this chapter has been the extent to which various parodic appropriations might be considered 'straight Shakespeare'. Ultimately, however, the question becomes nonsensical. As the explicit use of intertextual reference becomes increasingly widespread across all forms of artistic discourse, the line between what is 'parodic' and what is 'straight' begins to disappear. In any case, there is no such thing as a 'straight' production: every manifestation of a Shakespearean text in contemporary culture does something with it, and as Terence Hawkes has formulated it, 'Shakespeare doesn't mean: *we* mean *by* Shakespeare' (1992: 3). Parodic appropriation merely wears its cultural bias on its sleeve, so to speak, making the act of *acting* Shakespeare explicit: through its ironic distance, it draws attention to what Weimann might term its 'bifold authority', or Bakhtin its 'double-voicedness'. The persons *telling* us 'Shakespeare' are, in truly Brechtian fashion, always conspicuous, as are their attitudes towards the subject of their parody (dialogic and self-contradictory as those attitudes may be).

'Shakespeare' itself also has a fluid and constantly shifting set of cultural meanings, signifying frequently not so much the Shakespearean texts as the cultural values with which those texts are inscribed. As we have seen, these values are subject to change, constantly shifting on the basis of Shakespeare's use in education, politics – even in pop culture itself. Thus, for example, the 'Laurence Olivier impression' school of Shakespeare parody increasingly loses its currency, the further from his films the popular consciousness departs. Today, in fact, the audience's constructions of 'Shakespeare' might stem more frequently from *other* pop appropriations: from the films of Luhrmann and of Branagh, for example, or (as we shall see in Chapter 5) from the highly fictionalised depictions of Elizabethan theatre in *Shakespeare in Love* and *Doctor Who* – perhaps even from some of the parodic appropriations studied in this chapter. And as 'Shakespeare's' meanings shift, so too do the meanings of his pop appropriations. Always staging a dialogue – and often a confrontational one – between meanings old and new, popular appropriation of Shakespeare is ever engaged in a constant renegotiation over what, exactly, 'Shakespeare' means.

Personal Narrative 5
Blasphemy

The naked fat man shrieks and covers his genitals.

'It's an insult to women,' mutters the woman to my right, after the audience's laughter has died down.

The fat man is Ólafur Darri Ólafsson, and the context for his nudity is the Icelandic company Vesturport's production of *Romeo and Juliet*. It transferred to the Young Vic from the Reykjavik City Theatre in 2003, but I am watching it in January 2005 at the Playhouse Theatre in the West End. Ólafsson, a corpulent, bearded man, who has spent most of the performance in and out of a variety of multicoloured outfits, is playing the Nurse.

The show is provocatively blasphemous, both metaphorically and literally. One member of the company dons a loincloth and a crown of thorns, and stands in a crucifix position on a raised platform at the back on the stage; we hear another actor mutter 'Jesus Christ!' in a stage whisper, and the lights come up to reveal Friar Laurence's cell. The audience laugh, with some incredulity. The Jesus figure proceeds to light a joint (using a lighter he has tucked into his loincloth), and to share it with the Friar. Upon Romeo's entrance, both characters hurriedly extinguish their spliffs and assume more conventionally 'holy' poses.

An equally irreverent attitude is taken towards the sanctity of the Shakespearean text. In one scene, Erlendur Eiríksson's Capulet and Víkingur Kristjánsson's clownish Peter give a hilariously frantic rendition of 4.2.

'Marry, sir,' gabbles Peter, in a thick Icelandic accent, '''tis an ill cook that cannot lick his own fingers: therefore he that cannot lick his fingers goes not with me' (4.2.4–6). He turns to the audience and shrugs. 'I have no idea what I'm saying,' he admits, grinning. 'But *Shakespeare* wrote this, so ...'.

And on he soldiers.

Actors ad-lib with the audience throughout, and pop culture references abound. 'Luke, I am your father,' says Peter at one point, drawing an imaginary lightsaber; Paris bursts on stage singing 'I've Got You Under My Skin'. The cast deliver many of their lines in heavily accented English, but most of them lapse into Icelandic from time to time (particularly, I notice, at moments of great emotion). At less serious moments, Peter will run onto the stage and shout 'In English!' at them in mock exasperation.

The company enjoy the cultural differences that separate them from their audience, rather than (as I have seen in other intercultural performances) trying to play them down. Kristjánsson pretends to begin the play in his native language; 'Good Lord, Cynthia, they're going to do it in Icelandic!' he cries, aping an imaginary audience member.

'We Icelanders are famous for our delicious whale,' he later insists. 'And you English are famous for your ugly women!'

The audience boo, playfully.

'But,' he says, as the booing increases, 'but, but, but – that is certainly not the case tonight!'

An audience member arrives late, and he is hauled up on stage for ritual humiliation: he is, apparently, a German named Hans. Peter flirts with him a little (he flirts with male and female audience members alike) before asking him to read aloud the list of guests with which Capulet has left him. This comic sequence, of course, has roots in Shakespeare's 1.2, in which the illiterate Peter asks Romeo to read his letter. But Kristjánsson's Peter rapidly loses interest in his task, tricking Hans into an impromptu game of leapfrog, and pulling off his victim's sweater in the process.

Members of the audience gasp and begin to mutter. *One does not undress theatregoers in this country,* they seem to be saying. *Surely these Icelanders know that?*

But 'Hans', of course, turns out to be a plant. He cartwheels straight into the ensuing fight scene.

Despite all this, Vesturport's production is sincere in its iconoclasm. Romeo is played by director and former gymnastic champion Gísli Örn Gardarsson, and he and Nína Dögg Filippusdóttir's Juliet play many of their scenes together while hanging from trapezes. Their scenes are passionate, sexy, strangely moving. Their first kiss takes place in mid-air – Romeo hangs upside-down by his feet from a trapeze, and pulls Juliet up towards him by her hands. They remain suspended there, in a gravity-defying embrace, for several seconds. When Gardarsson's Romeo envisages his love as an angel who 'bestrides the lazy passing clouds / And sails upon the bosom of the air' (2.1.73–4), or explains to her that 'With love's light wings did I o'erperch these walls' (2.1.108), the images are literalised. Some of his stunts, which he performs gracefully and seemingly without effort, are met with gasps from the audience.

When we reach the end, the deaths are stirringly poignant. Romeo climbs a long, thin piece of white material which hangs from the ceiling. He delivers his final lines partly in Icelandic, with tears streaming down his face, and wraps the white silk around his feet. He lets himself drop, and dangles, mid-air, completely still.

Juliet's death follows exactly the same pattern as Romeo's: the climb, the tears, the Icelandic text, the drop. As she falls, four other members of the company perform an identical stunt behind the hanging couple.

I eavesdrop on audience conversations on the way out. The gentleman to my left seems moved and exhilarated; the ladies to my right, audibly outraged. A couple of teenage girls are enthusiastic.

The family sitting behind me, however, left during the interval.

'It's actually murdering Shakespeare,' I heard one of them say.

5
Shakespeare's Popular Audience: Reconstructions and Deconstructions

> [T]he Globe feels really *inclusive* rather than what I always hated about Shakespeare, that it was *exclusive* and for posh people. ... And this is what's made me want to do this programme, because I think that the thematic quality of Shakespeare should bring us all together. It speaks to all of us – working-class and kings.
>
> (Lenny Henry, *Radio Times*, 25–31 March 2006)

Defining a 'popular' audience

On Saturday, 25 March 2006, BBC Radio 4 broadcast a documentary entitled *Lenny and Will*, in which the comedian Lenny Henry explored his own fascination with, and intimidation by, Shakespearean performance, interviewing such theatrical illuminati as Sir Peter Hall, Dame Judi Dench, and Sir Trevor Nunn. Henry wrote a short piece on the broadcast for the same week's *Radio Times*; in a neat symbol of the documentary's characterisation of a clash between highbrow (Shakespearean theatre) and lowbrow (Henry the working-class comic), the article was accompanied by a photograph of a perplexed-looking Henry staring at a bust of Shakespeare. The quotation above is from that article, and I begin this chapter with it because it encapsulates some of the assumptions and aspirations which underpin many of the 'popular Shakespeares' considered in this chapter.

We are concerned with the 'popular audience'. Immediately we hit upon the problem of a definition. As we saw in Chapter 1, there are a variety of models – not all of them congruent – by which an audience might be understood as a 'popular' one. In what we termed a 'market' definition, the concept of a popular *audience* is relatively meaningless: it is simply a mass of actual or potential consumers of a popular *product*. Here, a popular audience is quite simply a *mass* audience. The 'anthropologically' popular audience, on the other hand, is very easily identified, since the anthropological definition takes the audience as its starting point: it is composed of 'the largest combinations of groupings possible' within its wider society (Mayer 1977: 263). In this sense, the popular audience is an actual social demographic, quantifiable and identifiable with reference to statistical breakdown (*x* many ABC1s, *x* many C2DEs, and so on).

We will return to such concrete definitions of the 'popular' in the next chapter; what this chapter considers, however, is a conception (or series of conceptions) of the popular audience which derives not from material fact, but rather from the perceived relationship between actor, play, and spectator – and crucially, between spectator and spectator. Chapter 1 described two other, ostensibly opposed, models of the 'popular', which I termed the 'collectivist' and the 'critical'. In the former, audience members are encouraged to establish or to strengthen a sense of communal identity; in the latter, such uncritical surrender to a group identity is discouraged. Both models tend to be championed by their proponents as politically progressive, but as we saw in the opening chapter, such assumptions are very much open to question. Central to the debate is the issue of shared response: the means by which the audience's sense of itself as a group is constructed.

Group response is not an easy matter to discuss; as we saw in the previous chapter, an impression of consensus is all there is to go on. Unlike demographics, it can never be accurately measured or quantified – it is always subjectively experienced, and while an audience may give the impression of 'an overall homogeneity of response' (Elam 2005: 87), a play is always, as Bernard Beckermann has put it, projecting doubly: 'to each member of the audience as an individual ... and to the audience as a whole, in that distinctive configuration that it has assumed for a particular occasion' (1970: 133). Try to unpick the latter, however, and as Peter Holland suggests, 'the audience fragments into its constituent individualities, dissolving the myth of a unity of reception and creating instead an unassimilable and unmeasurable diversity' (Holland 1997: 19). Questions of how group identity is constructed, the frames through which the audience experiences that shared identity, and indeed by *whom* it is shared, are central concerns of this chapter. What is the nature of the group constructed? How does that group stand in relation to the larger social and cultural groupings of which it is a part?

Shared experience: Unifying the audience

The Prologue to *Henry V* appeals to its audience for a kind of imaginative collusion: 'let us,' requests the Chorus, 'On your imaginary forces work. ... Piece out our imperfections with your thoughts' (17–23). When the speech was delivered at Shakespeare's Globe during the theatre's inaugural season in 1997, the audience's willingness to fulfil this performative role was, apparently, 'invariably' indicated by their audible, affirmative response to the Chorus' question, 'Can this cockpit hold the vasty fields of France?' (Kiernan 1999: 27).

A willing complicity of this sort, according to director Mike Alfreds, is what sets the theatre apart from film: actors create the imaginative world of the play in full acknowledgement of its fictional nature, and by implication, request the audience's complicity in establishing the illusion. Alfreds recounts how this realisation made him appreciate that actors and audience 'shared in an act of imagination – were, in fact, brought together by it' (Alfreds 1979: 4). As a result of their collaboration, he argues, the two groups achieve a powerful sense of togetherness; Alfreds founded the theatre company Shared Experience upon this very principle. It was seen in action in his production of *Cymbeline* for Shakespeare's Globe, when a cast of six played over thirty characters between them, transforming before the audience's eyes.

The ideal is a popular one in the Shakespearean theatre, and indeed in the theatre at large. David Selbourne describes the way in which an early performance of Peter Brook's *A Midsummer Night's Dream* established a 'warm and intimate relation with the spectators':

> 'Here's a marvellous convenient place for our rehearsal,' Quince had said to the mechanicals, with an eye on the audience. And when laughter of deepest complicity greeted him, a compact between actor and spectator – and between play and players, had been forged for the night, beyond breaking.
>
> (Selbourne 1982: 327)

Brook's 2000 production of *Hamlet* engendered similar 'complicity', with its cast of eight playing thirteen roles; Adrian Lester and Scott Handy came on, in Paul Taylor's words, as 'comically Riverdancing scene-shifters' before assuming the characters of Hamlet and Horatio 'in the graveyard they have just fashioned with cushions' (*Independent*, 8 December 2000). In Sidney Homan's production of *The Merry Wives of Windsor*, too, scene changes took place onstage, and the play ended with actors and audience in shared light (Alfreds advocates the same techniques; 1979: 10–13). In his book on the role of the audience in theatre, Homan describes the final moments of the production in similar terms to Alfreds': 'the now-solid Windsor community paralleled that united community of performer and audience who together had sustained the stage's illusory world' (Homan 1989: 162). Homan's book is titled *The Audience as Actor and Character*; as it implies, the audience of the collectivist theatre are expected to participate as co-creators of the play world.

For Alfreds, 'the actor's freedom to create and respond freshly at each performance' is one of the most important aspects of live theatre (1979: 5). In an interview with Clive Barker for *Theatre Quarterly*, for example, he explained that in his 1981 production of *Cymbeline*,

> [e]very night, in every show we do, no playing of a scene is fixed. ... Performances are a disciplined form of improvisation and still true to the text to the best of our ability and understanding.
>
> (Alfreds & Barker 1981: 22)

Again, it is a principle shared by many of those concerned with a more 'collaborative' Shakespearean theatre. In *Free Shakespeare*, John Russell Brown laments the primacy of director's theatre, in which the 'explorative and fluid engagement' for which Shakespeare's plays were written is denied:

> The director gives unity, the actors settle into their roles and the audience is kept in the dark to receive whatever view of the play has been chosen for them.
>
> (1974: 82)

Brown advocates a more actor- and audience-centred approach, in which actors are given only their own speaking parts, and perform after a minimum number of rehearsals ('sufficient only to arrange for the movement of supernumaries and the handling of properties'; 83) before a same-light audience, 'acting from a prepared knowledge and study of the text, and then improvising a response to the unfolding drama' (1974: 84).[1] Brown's proposed experiment has been borne out in practice most prominently by Patrick Tucker's Original Shakespeare Company, who have performed several times at Shakespeare's Globe; the company perform unrehearsed, as Tucker believes the Elizabethan actors to have done, having been given only their own lines and their cues. The result, explains Tucker, is that the actors experience the play at one with the audience: 'together they explore and enjoy the discoveries and ideas of the play' (2002: 41). Tucker reports the conclusion, after one such performance, of a member of his company: 'It is just Us and Us out there, not Us and Them' (2002: 41).

A focus on the virtuosity and indeed the very 'liveness' of performance keeps the play rooted in what Alfreds terms 'the dual nature of theatre' (1979: 5); one might recall Weimann's *locus* and *platea* or Counsell's 'abstract' and 'concrete' registers. Much like the accounts of improvisation discussed in Chapter 3, Tucker's description of his performance practice emphasises 'that essential danger that a live performance brings', and he constructs cue-script acting as 'very like a circus, or a sports event' (2002: 111). Many Shakespearean productions over the years have in fact literalised Tucker's analogy, importing performance modes from the circus in order to foreground the concrete register of performance. A recent trend seems to have been the use of aerial performance: notable examples include Vesturport's *Romeo and Juliet* (2003), Tim Supple's *A Midsummer Night's Dream* (2006), Michael Boyd's *Henry VI* trilogy (2006), and at the Globe, Kathryn Hunter's *Pericles* (2005) and Tim Carroll's *The Tempest* (2005); the reader's own theatregoing experience may well provide further examples. In emphasising the skill and physical risk involved in the creation of the performance, such productions draw attention to their own performativity. Gravity-defying stunts will frequently provoke audible responses from spectators – gasps, applause – prompting them to adopt patterns of behaviour more associated with sport than with theatre. This study has already considered the potentially carnivalising and deconstructive effects of circus on Shakespeare in its discussion of Vesturport and the Flying Karamazov Brothers; but the effect is not always to signify a disjunction from the text. As Counsell notes, an emphasis on theatre's duality can serve either to disrupt a stable reading of the play by offering *locus* and *platea* in contradiction, or to unify and to stabilise by showing them as interdependent (1996: 164). When characterised as an imaginative collaboration between actor and audience, circus-inflected Shakespeare implies the latter. Brook's *Dream* is, of course, the most famous example of this; Tim Supple's production of the same play might be considered another (Figure 5.1). Michael Billington's review was one of many which compared Supple's production, generally in highly favourable terms, with Brook's:

> [I]n its strangeness, sexuality, and communal joy this is the most life-enhancing production of Shakespeare's play since Peter Brook's ... the company finally joins in a candle-lit chant devised by the music

Figure 5.1 Company of Tim Supple and Dash Arts' *A Midsummer Night's Dream* (2006).

director, Devissaro. As in all the great *Dreams*, we feel we too have participated in an act of ritual communion.

(Guardian, 9 June 2006)

Though Supple's production was clearly not a 'faithful' one in a strictly textual sense – around half the text was spoken in languages including Hindi, Bengali, Marathi, Malayalam, Tamil, and Sinhalese – its unproblematic unification of its audience made gestures towards a Shakespearean universality which Vesturport's bilingual *Romeo and Juliet*, for example, did not. Unlike the press response to Vesturport's production, there were few complaints that Supple's *Dream* was 'not Shakespeare'.

This might be profitably explored in relation to the reading of the piece suggested by Billington's final sentence. Champions of 'collectivist' popular theatres frequently discuss their work in terms of ritual and communion; the most notable example is perhaps the influential Polish director Jerzy Grotowski, whose pared-down 'poor theatre' was intended to establish an 'actor-spectator relationship of perceptual, direct, live communion' (1969: 19). Grotowski's explanation fits very clearly the 'shared experience' model outlined above:

We found it was consummately theatrical for the actor to transform from type to type, character to character, silhouette to silhouette – while the audience watched – in a poor manner, using only his own body and craft. ... By his controlled use of gesture the actor transforms the floor into a sea, a table into a confessional, a piece of iron into an inanimate partner, etc.

(1969: 20–1)

Having collaborated in a process of imaginative communion, the spectator's emotional response will ideally become similarly unified with that of the rest of the audience; 'communal theatre,' explains Philip Auslander, 'carries artists and audience together to a level of universal emotional response then returns them to quotidian reality with a keener sense of the psychic structures shared by all people' (1997: 19). Ideally, one achieves a sort of catharsis, but a catharsis which is all the more powerfully experienced because it is experienced in a communal context, in which others appear to be experiencing the same. Peter Brook writes of the need for

a circle of unique intensity in which barriers can be broken and the invisible become real. Then public truth and private truth become inseparable parts of the same essential experience.

(1989: 41)

Such aspirations are, however, as Auslander notes, 'based on disputable psychological and semiological assumptions': namely belief in the existence of a universally shared collective unconscious (or something similar), in the possibility of a genuinely unified communal psychological response, and the conviction that art 'can have a direct effect on its audience at the psychic level, and that this effect, however it is defined, is ultimately of a health-giving nature' (1997: 27).

Though Brook found many 'parallels and points of contact' between Grotowski's theatre and his own (1989: 38), Grotowski's work focused increasingly upon the self-revelation of the actor, leading ultimately to a 'paratheatre' which negated the need for an audience at all. Brook's theatre, on the other hand, remained firmly collectivist, 'leading out of loneliness to a perception that is heightened because it is shared' (1989: 41): he described his work in Paris as an attempt to 'reunite the community, in all its diversity, within the same shared experience' (1978: 7, quoted Williams 1985: 41). For Brook, there are two contrasting means by which an audience might display such unity:

There is the climax of celebration in which our participation explodes in stamping and cheering, shouts of hurrah and the roar of hands, or else, at the other end of the stick, the climax of silence – another form of recognition and appreciation for an experience shared.

(1990: 52)

The theatre of noise and applause, Brook called the 'Rough'; that characterised by 'deep and true theatrical silence', he termed the 'Holy' (1990: 76, 49). It is, however, when Rough and Holy combine that theatre for Brook is at its most immediate – and its most 'Shakespearean':

> It is through the unreconciled opposition of Rough and Holy, through an atonal screech of absolutely unsympathetic keys that we get the disturbing and the unforgettable impressions of his plays. It is because the contradictions are so strong that they burn on us so deeply.
>
> (1990: 96)

An emphasis on both public and private, Rough and Holy, is what enables Shakespeare's plays to 'present man simultaneously in all his aspects':

> [T]ouch for touch, we can identify and withdraw. A primitive situation disturbs us in our subconscious; our intelligence watches, comments, philosophizes. ... We identify emotionally, subjectively – and yet at one and the same time we evaluate politically, objectively in relation to society.
>
> (1990: 98)

The audience engagement which Brook expresses here is more complex than the uncomplicated 'togetherness' implied by more simplistic models of 'collectivist' theatre. Though both applause and silence are, for Brook, displays of audience unity, they are of completely opposite types: the interplay they entail between concrete and abstract, text and metatext, implies a dialogic theatre closer to Weimann's understanding of Elizabethan drama, and, of course, to Brecht ('Everything remarkable in Brecht', Brook once commented, 'is in Shakespeare'; 1989: 54). It should be noted, however, that Brook's assumption of Shakespeare's universality contradicts his claims for an 'objectively' Brechtian evaluation. For Brecht, concrete, *platea*-like playing was a defamiliarising device; for Brook, it is a simply different type of 'familiarisitation' with the play world.

The effects of the collectivist Shakespearean theatre in practice may be analysed with reference to the theatre work at Shakespeare's Globe, where the house style might be seen to epitomise many of the key tenets of a 'Shared Experience' style. Indeed, Mark Rylance worked with Shared Experience at the National Theatre in 1987, and Mike Alfreds directed at the Globe twice during Rylance's tenure as Artistic Director; Rylance has cited Alfreds as a key influence on his own theatre practice (Rylance 1997: 171). At the Globe, actors and audience necessarily share the same light; imaginative complicity is essential for the creation of illusion (Rylance would thank the audience for their 'imaginative generosity' at the end of *The Tempest* in 2005); as we have seen, there have also been experiments there in emphasising the physical virtuosity of performance. Directing practice there (particularly under Rylance's artistic directorship) has often followed the broad principles discussed above: it is frequently called an 'actor's theatre', and actors have often been encouraged by directors to create anew every performance.[2] Most conspicuously, the audience is an active one: Pauline Kiernan reports that laughter, applause, and other vocal responses at the Globe have 'often lengthened performance time by as much as 15 minutes' (1999: 123), and several Globe actors have drawn comparisons with sports audiences (Rylance 1997: 171; Kiernan 1999: 22, 134, 149). In an illustrative anecdote, Kiernan recounts the story of a playgoer at *The Two Gentlemen of Verona* (1996) who cried out 'Don't do it, Julia!' as the character was embraced by Proteus, suggesting a potentially progressive audience agency (1999: 33).

Frequently, it has been suggested, audience response at the Globe threatens to undermine the effectiveness of the performance. The theatre has gained what *The Stage* described in 2001 as 'something of a reputation for audience participation' (24 May 2001); in 1998, Benedict Nightingale issued a 'friendly warning' to Rylance in the pages of *The Times*, questioning whether such participation would spoil the work there 'in ways that rainstorms and passing helicopters have failed to do':

> Are we *en route* to the *Hamlet* in which Claudius makes V-signs at the groundlings, and the groundlings yell 'Go for it, boy' at the hesitant prince? ... An audience which jeered Iago or yelled 'Look out, he's lying' to his victim would doubtless be an involved audience; but they would be unlikely to end up with a very searching, subtle *Othello*.
>
> (*Times*, 3 August 1998)

The following year, Alastair Macaulay described the Globe as 'without philosophy, without tragedy, without elegance, without power, without stillness' (*Financial Times*, 28 May 1999) – in Brook's terms, perhaps, the 'Rough' had overwhelmed the 'Holy', causing a fatal imbalance. But others have suggested a more optimistic outcome for the Globe's experiment, finding potential in the theatre's transition from shared noise to shared silence which evokes Brook's ideally 'Shakespearean' combination of Rough and Holy. In an *Evening Standard* interview, Mark Rylance described how he had discovered at the Globe that Shakespeare 'opens people's hearts with humour and then plunges the knife in':

> I've noticed the wit in Shakespeare is much more obvious here, particularly in the tragedies. ... When Cleopatra says: 'How heavy weighs my lord' as she lifts him just before he dies, we get a laugh that I hadn't expected. Then, within moments, she's howling the most terrible pain and the audience is suddenly, absolutely, silent.
>
> (31 July 1999)

John Peter noted in the Globe's opening season that during such moments, 'a silence descends, an almost palpable silence of hundreds of attentive people close together, such as I have not experienced before' (*Sunday Times*, 15 June 1997). In many ways, this phenomenon is the antithesis of audience play-acting; but as Brook's model suggests, this moment of communion might be characterised as a direct result of the togetherness established during moments of audience activity. 'At moments when the audience is emotionally moved and goes quiet,' explains Kiernan, 'its participation is as palpable in its still silence as in its animated noisiness' (1999: 36).

Idealism and nostalgia

Models of theatrical 'shared experience' often centre around notions of equality, and many of its advocates discussed above describe the politics of such work in idealistic but somewhat unspecific terms. 'Theatre,' says Alfreds, 'is the occasion on which we confirm our shared humanity' (1979: 5); Homan describes the 'ethical dimension of the theatre' as 'a bringing together, a joining of two groups, one onstage, the other offstage' (1989: 164). Describing the implications of a Jatra performance he witnessed in India, in which actors were not 'segregated' from their audience, Brown concludes that 'it was political because, imaginatively, they were all in it together' (Brown 1999: 25). At the Globe, meanwhile, Rylance argues that the audience's voice is 'liberated and powerful', since 'actors play and sometimes talk directly *with* the audience, rather than *to* or *at* them' (*Times*, 14 August 1998). Watching the Globe's 1998 production of *As You Like It*, Peter found it 'impossible not to experience a sense of togetherness, of shared enjoyment that cuts across the price of tickets': noting that 'even the learned jokes can be angled so that the unlearned, too, can laugh at them', he speculated that 'perhaps a shared culture is not such an impossibility after all' (*Sunday Times*, 7 June 1998).

Though they may be rooted in the progressive desire for a greater equality, the politics of such statements must, however, be questioned. Firstly, the impression of actor/audience equality is rarely more than a semblance of it: this view, as Auslander argues, 'misunderstands the dynamic of performance, which is predicated on the distinction between performers and spectators' (1999: 56). Even when the audience are active participants in the play's action, their participation is generally controlled and structured by cues from the performers. Secondly, claims of equality among the audience themselves must be subject to serious scepticism. A 'shared culture' can never be a culture without politics; a 'shared humanity' will always be shaped in the image of *somebody's* version of humanity. When a unified audience is invoked, then, one must question *who* is doing the shaping: upon whose image is the unified audience being modelled? Where do the cultural affiliations of this Utopia lie?

Such desires often find themselves looking backwards to an imagined culture in which audiences experienced none of the fragmentation and alienation of contemporary life. Brook described his desire for a 'collective experience' shared between actors and audience as 'a new Elizabethan relationship' (1989: 40), and indeed it is most often the Elizabethan theatre upon which such Utopian nostalgia is modelled; it is, after all, the Elizabethan theatre which is invested with Shakespeare's authority. An 'authentic' popular Shakespeare is not, of course, completely unfounded in historical fact, and the evidence which does exist regarding the Elizabethan theatre and its audiences has led respected theatre historians to write of 'the democracy of the audience' in the Elizabethan playhouse (Harbage 1969: 12), or to conclude that Shakespeare's audience 'expressed their pleasure and sense of discovery by instantaneous applause' and that his plays 'were written for this immediate, shared sense of discovery and achievement' (Brown 1974: 80–1). Such accounts are naturally influential in shaping contemporary practice which aims at Shakespearean 'authenticity'. Andrew Gurr, for example, writes:

Crowds strengthen their sense of identity, their collective spirit, by vocal expression of their shared feelings. The audience was an active participant in the collective experience of playgoing, and was not in the habit of keeping its reactions private.

(2002: 46)

It is not difficult to imagine that, as the chief academic advisor in the reconstruction of Shakespeare's Globe, his words might have had a formative effect upon theatre practice there.

The scholars cited above are, of course, too diligent to make erroneous claims for an authentically 'shared' Elizabethan culture in which class differences played no part. Others, however, have been more selective in their evocations of history. When John Peter constructs Shakespeare's theatre as a 'living tradition that was serious, entertaining, commercially viable and classless' and in which 'the sense of social differences between the groundlings and the seated spectators would have been minimal' (*Sunday Times*, 15 June 1997), he is perpetuating a myth which has an illustrious pedigree in literary criticism. E. M. W. Tillyard's 1943 book *The Elizabethan World Picture* (according to Tom Matheson, 'once the almost canonical account of commonly held Elizabethan beliefs'; Dobson & Wells 2001: 474) argued for a shared Elizabethan belief system which was 'so taken for granted, so much part of the collective mind of the people, that it is hardly mentioned except in explicitly didactic passages' (Tillyard 1990: 17). In *The Death of Tragedy*, George Steiner suggested that Shakespeare's audience must have been 'a community of men living together ... in a state of tacit agreement on what the nature and meaning of human existence is' (1961: 113).

The past is, by its nature, irretrievable. As we imagine it, we refashion it in our own image – or in the image of our own desires. To point out that such Utopian nostalgia misrepresents the past, writing out the realities of class difference and of exploitative power structures, is to state the obvious; what does need articulation, however, is a sense of the political implications of such mythologising for the present. Upon closer inspection, the modern fantasy of a 'universal' Elizabethan audience, founded as it is upon an essential 'Englishness' and a longing for a return to a more stable past, serves interests suspiciously close to the political hegemony. It may be worthwhile to ask not only which groups (and their interests) are included in the imagined unification, but also which are excluded, since inclusion cannot take place without exclusion. The 'community' established will define itself in relation to politics of race, nationality, class, and gender, and if these politics are obscured by reference to 'universal humanity', the more likely it is that they are fundamentally conservative. The politics of nostalgia are always specific, and this comes into sharpest focus when specific examples are considered. At the Globe in 1997, when the apparently unified ground-lings were addressed by Rylance's Henry V as if they were his soldiers at Agincourt, they were not merely participating in the creation of the play world, bound together by Shakespeare's universality; they were sharing in a nationalistic fantasy of Englishness within a particular social and politi-cal context, shaped, for example, by the recent defeat of the Conservative Party by New Labour, or by the 'handover' of Hong Kong to China which took place that July. The following year, Benedict Nightingale's 'friendly warning' expressed a deep worry about what he saw as the 'self conscious, phoney role-playing' of the Globe's audience:

> [H]ow does he [Rylance] feel when he stands in the Doge's court as Bassanio in the current *Merchant of Venice*, and hears Shylock booed and his forced conversion to Christianity received with the same cheers that have apparently sometimes greeted the grossly anti-Semitic Gratiano? ... Last year it was much the same. The audience made its feelings about the Agincourt campaign very evident. The dastardly Gauls were booed and even Henry V's decision to kill his prisoners cheered.
>
> (3 August 1998)

Rylance's reply in the same newspaper denied Nightingale's assertion that the killing of the prisoners was ever cheered, and constructed the responses to Shylock as 'not anti-Semitic but disapproval of his murder-ous intent'. Like Nightingale, Rylance expressed concern about 'hissing and booing', but blamed this partly on 'inaccurate press material' which had apparently misled the audience as to their required response (*Times*, 14 August 1998). That Rylance was able to blame the media at all for the responses of his audiences, it should be noted, indicates that the models for their behaviour do not stem merely from the Globe's 'authentic' theatre practice and from its distinctive configuration of space.

Even when the gathered public appears to transcend such barriers of nationality and culture, similar problems are not avoided. Herbert Blau argues that when we are nostalgic about the 'cultural unity of audiences in the past',

> it may be well to remember the materialist detachment with which Walter Benjamin assessed, in his theses on history, the price of that

(seeming) unity and its 'cultural treasures.' For without exception, he wrote, they have an origin in exclusion and exploitation that cannot be thought of without horror.

(1990: 29)

He goes on to suggest that such analysis is true not only of nostalgia for the imagined unity of the Elizabethan audiences, but also of audiences in India and Java today, 'which we still imagine as fulfilling, on largely unremembered premises, the unifying function of the recurring communal dream' (1990: 29).

As Blau indicates, intercultural performance complicates nostalgic longing even further. Writing about Brook's *Hamlet* for *Shakespeare Survey*, Michael Dobson found himself cynical about the production's 'quest for universal spiritual truth' – an aspiration which he felt made it seem 'like a belated twentieth-century survival into a less optimistic era':

Brook's production sought in *Hamlet* for a set of pan-religious spiritual truths supposedly accessible to all cultures at once, attempting to set the play timelessly and everywhere. The First Player was Japanese, the Gravedigger Irish, Ophelia Indian; the music was predominantly East Asian, the principal set an orange Persian carpet.

(2002: 302)

In the same article, Dobson discussed Alfreds' *Cymbeline* at the Globe as a production which 'seemed to have picked up a number of its tics from Brook':

[H]ere again, for example, were the onstage accompanists commenting on the action on assorted Asian instruments (here mainly percussion), and here again was the single carpet used to define the acting area – orientalist to the point of self-parody, as if Kai Lung had just unrolled his mat.

(2002: 314)

As Dobson's accounts imply, such gestures towards supposedly acultural universality are simply that – gestures. When productions such as Brook's or Alfreds' prime their audiences, through their use of culturally eclectic staging conventions, to look for Shakespeare's universal meanings, they may in fact be constructing the 'universal' in the image of the West. Counsell describes Brook's work since *A Midsummer Night's Dream* as pervaded by an 'aura of eastern-ness' (1996: 170); in *Hamlet*, indeed, several cast members were drawn from a variety of Asian cultures, and the design – burnt-orange carpets, candles, tunics, sitar – was unspecifically 'Eastern' in its composition (Robert Hewison described the costumes as 'artfully indeterminate'; *Sunday Times*, 26 August 2001). Rustom Bharucha criticises such work on the grounds that 'it exemplifies a particular kind of western representation which negates the non-western context of its borrowing', and suggests that Brook uses India 'as a construct, a cluster of oriental images suggesting timelessness, mystery and eternal wisdom' (1993: 70, 83).[3] As Counsell points out, such mystical associations are 'highly suspect', and reduce 'the complexity of many cultures to a single figure of *western* iconography' (1996: 171). Such romanticisation serves to obfuscate the very forces of western materialism which keep the 'East' oppressed: by imagining that the spiritual communalism of such societies enables them to transcend poverty, western audiences let themselves, and their own cultures, off the hook.

Such analysis implies that nostalgia for a spiritually richer past obscures clear sight of power relations in the present. Proponents of the collectivist theatre frequently posit a decline in spirituality as the source of contemporary disaffection in the West; Brook, for example, argues that

> [w]e try to believe that family bonds are natural and close our eyes to the fact that they have to be nourished and sustained by spiritual energies. With the disappearance of living ceremony, with rituals empty or dead, no current flows from individual to individual and the sick social body cannot be healed.
>
> (1989: 136)

In a similar vein, Rylance looks back to a specifically Elizabethan past in which answers might be found:

> Our world view today forms a very different picture from the Elizabethan. Many now scorn any suggestion of the influence of the heavens on human activity. Modern psychology has tried, with no more than limited success, to interpret and cure the soul. As a classical actor, I naturally have enormous faith in classical drama as an enriching and beneficial force in the individual and society.
>
> (Rylance 1997: 175)

Rylance's extra-theatrical political engagement is hardly to be doubted: during his time as Artistic Director of the Globe, he worked with Peace

Direct, the World Development Movement, and the Campaign Against Arms Trade. In the theatre itself, however, he has been slightly more ambivalent. When asked in an interview with the 'health and healing' magazine *Caduceus* whether he was moving his work 'into a more political arena', he replied:

> We need more people who promote conciliation and honouring of both sides, who question the means by which they are resolving their issues. In a global village the time has passed for violent resolution of issues. ... Therefore when you ask whether I want to be more political, my answer is that I do, but I don't want to be just another angry person.
>
> (Peterson 2005: 9)

The role of the theatre, he concluded, was to 'provide some ritualistic movement to the year, primarily to help people to get a soul's view of what is happening' (2005: 9). Certainly, like the critical attitude advocated by Brecht, this implies a theatre which ruptures audiences from the everyday attitudes they take for granted; whether it provides an 'unillusioned' one, however, must be open to question.[4]

Simulating the popular Shakespearean audience

It may be worthwhile to consider, for a moment, how the image (illusion?) of popular communality is created in Shakespearean theatre. As the deconstructions above suggest, it might be better understood as a *simulacrum* of *communitas*: an idealised *copy* of a phenomenon which never really existed in the first place; or in Blau's terms, 'the merest facsimile of remembered community, paying its respects ... to the better-forgotten remains of the most exhausted illusions' (1990: 1). Audience behaviour is never an 'organic' response to a performance, unmediated by codes and conventions of spectatorship learned from culture at large; as we saw in Rylance's own defence of the Globe, it may in fact be an emphatically 'mediatised' response. W. B. Worthen notes that film and television play a large part in shaping what he calls the audience's 'performativity' at the Globe, comparing their 'habits of participation' to those learned at theme parks (2003: 86): in a sense, he argues, 'what the patrons "perform" in theme parks is experience troped by the imagery, characters, and narratives previously encountered in mass-culture entertainment' (2003: 94). I should like to extend this analysis a little further, to suggest that in some ways, the 'habits of participation' encouraged by popular Shakespearean performance at large may perhaps be similarly inflected with the 'tropes' of Shakespearean spectatorship learned from, and rehearsed in, the contemporary mass media.

The influence of mass entertainment in shaping popular attitudes towards Shakespeare should not be underestimated. *Shakespeare in Love*, for example, is widely used in secondary education despite its many evident historical inaccuracies, and a sense that Globe audiences often model their experiences at the theatre upon the film's depiction of Elizabethan performance may be inferred from the way in which the theatre's tour guides make frequent references to it.[5] One should be wary of ascribing too much influence to such media representations, however: mass-entertainment depictions of Shakespearean spectatorship are generally more symptomatic of prevailing cultural attitudes than they are paradigm-setting. But popular accounts of history are continuously evolving, taking new angles on popular myths, letting slip aspects of the old. As Richard Schechner has argued, history in performance is never 'what happened', but rather 'what is encoded and transmitted':

> [T]he event to be restored is either forgotten, never was, or is overlaid with other material, so much so that its historicity is irrelevant. ... Performance is not merely a selection from data arranged and interpreted; it is behaviour itself and carries with it a kernel of originality, making it the subject for further interpretation.
>
> (1982: 43)

Thus, for example, audiences may come to the Globe in 2007 having learned behaviours deduced by Alfred Harbage's historical analysis in *Shakespeare's Audience* (1941), but which were transmitted by Laurence Olivier's *Henry V* (1944), reshaped in *Shakespeare in Love* (1998),[6] and re-enacted at the turn of the twenty-first century at the Globe, which was itself the location for a highly fictionalised historical reimagining of Shakespearean spectatorship in a 2007 episode of *Doctor Who*.

Like all myths, the Globe's is one which is handed down from generation to generation, remoulded in the shape of contemporary concerns – parts are adapted, others discarded – and it may be worth tracking the evolving mediatisation of this myth with reference to the media representations mentioned above before we arrive at a broader consideration of contemporary media myths of Shakespearean spectatorship. Olivier's film, as Lanier puts it, 'has become a defining statement of the popular myth of the Globe' constructing the theatre 'as the site of an idealised, democratic popular culture' (2002: 145, 144). The audience's shared, often patriotic responses to the performance, and the lack of evident social divisions between them, recall Harbage's account of the 'democracy' of the playhouse; as Russell Jackson puts it, '[i]n what was at the time a powerful contrast to the drab, war-damaged London of 1944, Shakespeare's city is shown as idyllic, and its theatre as ideally democratic, colourful and lively' (2000: 27).[7] When elements of the film are recycled by John Madden in *Shakespeare in Love*, it is often to emphasise such class-transcendence. With its pan across a recreated Elizabethan London and close-up of a playbill, the later film's opening is, as Carol Chillington Rutter observes, both a quotation and subversion of Olivier's (2007: 249–51): unlike the earlier film's idyllic, picturesque Bankside, though, Madden's is dirty, littered, and lived-in (even the playbill is a piece of discarded litter). The film's characters are 'people like us': its anachronistic register allows them to speak in more-or-less twentieth-century idioms, and when we reach the performance of *Romeo and Juliet* which forms the film's climax, we know members of both the 'cast' (Will, Viola, Alleyn, Fennyman, Wabash) and the 'audience' (Henslowe, Burbage, Nurse), sharing their anxieties as the performance is jeopardised, and revelling in its eventual triumph.

The sense of the social divisions among the Elizabethan audience is greater than in Olivier's film: prostitutes and urchins mingle in the pit, while aristocrats like the villainous Lord Wessex and (in a flight of fancy) even the disguised Queen Elizabeth sit in the galleries. This serves only to emphasise the unifying effects of the play, however, since Madden's film, unlike Olivier's, depicts the audience as undergoing a profound transformation. Madden reveals on his DVD commentary track that he tried to maximise the amount of time he could spend filming with a theatre full of extras; he framed as many shots as he could from the actors' perspectives, so that the cinema audience could observe the faces of their Elizabethan doubles as they became 'held and captivated by what they were seeing'. In this film, the theatre audience are at first highly vocal, cheering and gasping, for example, during the play's fight sequence (Madden encouraged the extras to think of it as analogous to a football or boxing match). As the play reaches its climax, however, they undergo a transition from rowdiness to silence reminiscent of Brook's ideal combination of 'Rough' and 'Holy' modes of spectatorship.[8] The effect is transforming. The play ends, and as the screenplay indicates, '*There is complete silence. The ACTORS are worried. But then the audience goes mad with applause*' (Norman & Stoppard 1998: 145). Spectators of all classes are united in a standing ovation; even the Puritan preacher who had railed against the theatre earlier in the film is 'converted', applauding wildly and weeping.

Many of these elements are recycled once again in the 2007 *Doctor Who* episode 'The Shakespeare Code', though since the series is a time-travelling science fiction show, this is at a much further ironic remove from any claims to historical 'authenticity' (the Doctor enlists Shakespeare to help him defeat a trio of witch-like aliens called Carrionites). The myth, however, is still very much in evidence. Our first view of the Globe is at the end of a performance of *Love's Labour's Lost*: the audience is unified in cheering, and even those in the galleries are standing to applaud. What *Shakespeare in Love* implied through its encouraged identification of the cinema audience with their screen counterparts, 'The Shakespeare Code' makes explicit: the Doctor's companion Martha Jones (essentially the twenty-first-century television audience's onscreen representative in the Season Three *Doctor Who* adventures) is every bit as enthusiastic as her Elizabethan fellow-playgoers, deeming the performance 'just amazing!' A contemporary audience, the episode implies, might experience exactly the same sort of togetherness as the Elizabethans apparently did through watching Shakespearean performance.[9] Motifs from *Shakespeare in Love* are repeated throughout: Shakespeare is once again shown as a literary magpie, hearing many of the lines he will eventually write spoken first by other characters; the Puritan preacher resurfaces ('I told thee!' he gloats, as the Carrionites attack the playhouse); the climax once again features ecstatic applause and an appearance at the playhouse by Queen Elizabeth.

Both *Shakespeare in Love* and 'The Shakespeare Code' subscribe very firmly to contemporary myths of the 'popular' Shakespearean audience in mainstream popular culture. These myths are, it should be noted, a distinctly modern phenomenon. Prior to what Richard Burt has called 'the millennial resurgence of Shakespeare on film and television that began with Branagh's *Henry V'* (Burt & Boose 2003: 1), popular culture's depictions of modern Shakespearean spectatorship were typically far from collectivist: Shakespeare audiences were generally portrayed as distinctly upper-class, sitting in evening dress in darkened rows, often to be disrupted by intrusions from an adversarial popular culture (see, for example, the 1980s TV advertisement for Carling Black Label discussed in Longhurst 1988: 67–8, or the interrupted performance of *Othello* at the climax of the 1991 comedy *True Identity*). The following analysis of what is, I hope, a representative sample of recent depictions of Shakespearean audiences (both historical and contemporary) within popular culture, will show that the myth of a genuinely popular Shakespeare has gained in currency.

Mark Thornton Burnett identifies an emergence over the last 20 years of 'filmic representations whose narratives prioritise theatrical shows and stagings of Shakespearean texts' (Burnett 2007: 7).[10] My brief analysis will cover seven such representations. In *Dead Poets Society* (dir. Peter Weir, 1989), Neil Perry, a student at a strict boys' school in 1950s Vermont, disobeys his father to take the role of Puck in a local amateur production of *A Midsummer Night's Dream*. Kenneth Branagh's comedy *In the Bleak Midwinter* (1995) charts the troubled rehearsal process of a motley group of actors as they prepare for a Christmas performance of *Hamlet* in a church hall. The titular hero of *The Postman* (dir. Kevin Costner, 1997) travels across a post-apocalyptic America performing his one-man (and one-horse) version of *Macbeth*. American teen comedy *Get Over It* (dir. Tommy O'Haver, 2001) centres around a high-school musical adaptation of *A Midsummer Night's Dream* entitled *A Midsummer Night's Rockin' Eve*. *Stage Beauty* (dir. Richard Eyre, 2004), set in 1660s London, climaxes in an explosive performance of the final scene of *Othello* by its two central characters. *Shakespeare in Love* and 'The Shakespeare Code' have already been discussed.

Though the tone and setting of each of these films differ enormously, all of them construct their performances of Shakespeare's plays in a comparable fashion. The most striking correspondence is that they generally emphasise their fictional audiences' unity: spectators are shown to be vocal in their appreciation of the plays, cheering and applauding, frequently displaying their communality through simultaneous gasps, laughs, or groans. Often, the camera will pan across faces in the audience, calling attention to the homogeneity of the playgoers' responses. In all but one of these films (*The Postman* being the exception), the performance of the Shakespearean play forms an integral part of the climax, and when this is the case, the performance invariably culminates in a standing ovation. In most cases, the ovation unfolds slowly, as one or two playgoers stand up first; then the soundtrack swells, and in a rush of collective solidarity reminiscent of the famous denouement to Kubrick's *Spartacus* (1960), the rest of the audience stand to join them.[11] The implication, of course, is that the audience has found a collective voice through their shared identification with the Shakespearean performance. In the case of Branagh's film, in fact, the actors are explicitly concerned with the revitalisation of the local village community via the rescue of its church: 'We're not just doing a play,' says Joe, the director; 'We're here on a mission. To save this place. To get the developer out and the people back in' (Branagh 1995: 15). Later, Joe's sister Molly expresses it in even clearer terms:

> [T]here's no village hall, there's no arts centre. I mean, we need this place to give people a focus. Prove to the council that there is a community worth maintaining. That there *is* a community.
>
> (Branagh 1995: 22)

The play's rapturous reception at the film's conclusion suggests that such communality has indeed been proved, with Shakespearean performance as its catalyst.

The modes of performance enacted in these films have much in common with the ideals of collectivist theatre expressed above, and Shakespeare is characterised as 'genuinely' popular. In stark opposition to the pre-1990s cliché discussed above, all of the performances take place outside of professional proscenium arch theatres. In most of the films, audiences are shown to be composed of a wide variety of classes and ages (though notably not always races). The young are often foregrounded: *Midwinter* and *The Postman* both feature close-ups of (in Branagh's own words) *'riveted children's faces'* (1995: 74), while *Dead Poets Society*, set as it is at a boys' school, naturally emphasises the effects of the play upon its young protagonists. Shakespeare, according to these films, is potentially accessible to everyone. In *Get Over It*, in fact, it is the snobbish teacher Dr Desmond Forrest-Oates who does *not* 'get' Shakespeare; Forrest-Oates displays his philistinism throughout the film ('Bill Shakespeare is a wonderful poet, but Burt Bacharach he ain't'), while the voice of the 'true' popular Shakespearean is provided by the good-natured student Kelly ('It's kind of hard to understand, but once you get into it, the story's kind of good'). *Get Over It*'s insistence on the *story* as the essence of Shakespeare's play is echoed in its omission of virtually all the Shakespearean text from its eventual performance of *A Midsummer Night's Rockin' Eve* in favour of songs, dance routines, and adapted or improvised dialogue, and thereby lends 'Shakespearean' authority to the subgenre of which the film itself is a part: the high-school Shakespeare adaptation.

Indeed, many of the films emphasise spectacular elements over the Shakespearean text. Onscreen audiences are amazed by an 'improvised' dance in *Get Over It*, enchanted by a movement sequence in *Dead Poets*, terrified by a loud machine gun in *Midwinter*, and awed by special effects (rudimentary in *The Postman*, pyrotechnic in *Get Over It*, and genuinely extraterrestrial in *Doctor Who*). In several of the films, fight sequences elicit the most vocal responses (*Shakespeare in Love*, *Midwinter*, *Stage Beauty*). Ironically, given that these films are both metafictional and recorded, they tend to place great importance on the 'real' and on the 'live'. This often takes the form of an intrusion from the 'real' world: in *Get Over It*, Berke alters Lysander's lines to express his love for Kelly; in *Midwinter* and *Shakespeare in Love* respectively, Joe and Viola surprise even their fellow actors when they enter the performance. Characters undergo touch-and-go personal triumphs before our eyes: Kelly gets to perform the song she has written (*Get Over It*); Wabash overcomes his stutter at the last minute (*Shakespeare in Love*); Neil proves his abilities as an actor (*Dead Poets*).

Performances are invested with 'genuine' emotion from the actor-characters' offstage lives: Kelly's unrequited love as Helena, Neil's plea for forgiveness as Puck, Nina's fury as Ophelia (*Midwinter*), Will and Viola's passion as Romeo and Juliet (*Shakespeare in Love*), and Kynaston's anger as Othello (*Stage Beauty*), are all expressions of the characters' 'real' emotions. 'Don't act with what isn't there!' barks Kynaston to his co-star; Branagh's *Midwinter* screenplay stipulates that the actors' performances are '*utterly real*' (1995: 77), and Norman and Stoppard's, '*intense and real*' (1998: 139). Both *Shakespeare in Love* and *Stage Beauty*, in fact, set long before the development of naturalistic theatre, depict anachronistic triumphs of realism. Perhaps both performances-within-performances serve, as in Lanier's analysis of the play-within-a-play of Hoffman's *A Midsummer Night's Dream* (1999), as 'a legitimation of cinematic naturalism' (2003: 160).

The constructed nature of these characterisations of a popular Shakespearean audience becomes all-the-more evident when they fail to ring true. *Midwinter's* cheering and gasping crowd seems particularly staged (especially when, as Nina's Ophelia slaps Joe's Hamlet, they emit a communal 'ooh!'), and the film never quite explains why many of them feel free to break with theatregoing convention and stand and cheer during Hamlet and Laretes' duel. The *Dead Poets Society* 'Dream' provokes apparently spontaneous laughter from its audience in some very odd places, and Neil's frankly limp performance as Puck barely justifies the response he gets from the audience ('He's good. ... He's really good'), or the way in which his fellow cast members push him forward at the end to take an individual 'star' bow. Most unconvincing is the response of the audience to the rough-theatre *Macbeth* in *The Postman*, when the advance of Birnam Wood elicits joyous laughter from the children, and Kevin Costner's garbled delivery of Macbeth's most nihilistic soliloquy ('*Our* brief candle ...') appears, inexplicably, to inspire expressions of rapture. The actual content of the Shakespearean text is clearly not what matters in these instances; 'Shakespeare' becomes merely the token for the unifying and inspirational power of theatre, divorced from any literal meanings the text may have.[12]

These films play out – to varying degrees of credibility – the fantasy of a unified, popular Shakespearean audience, expressing the desire for an 'authentic' experience which is characterised by liveness and communality. It may be, as Auslander points out, that mass media itself has created the currency for such a desire: 'like liveness itself,' he argues, 'the desire for live experiences is a product of mediatisation' (1999: 55). Indeed, many of the practitioners examined in the first part of this chapter construct the 'liveness' of their theatre in direct opposition to what Alfreds calls 'the "fixed" perfection of film' (Alfreds 1979: 4; see also Grotowski 1969: 18–19, and Tucker 2002: 110–11). What was once an aspiration of the theatrical avant-garde has, it seems, now passed into the mainstream and been adopted by popular culture at large. The fantasies propagated by the films cited above play before cinema audiences which, if the clichés are to be believed, are typically *dis*-unified (though as Auslander points out, distinctions between theatre and cinema audiences are by no means this clear-cut), or before individual viewers watching in the privacy of their own homes. In essence, they are mere simulacra of a supposedly 'authentic' Shakespearean *communitas*, deriving their effect from the perceived absence of *communitas* among the mass popular audience.

A more nuanced politics of *communitas*

When the audience of these mediatised fantasies of popular Shakespearean theatre attend a live performance, then – particularly if, as at the Globe, it is one which is modelled on the collectivist ideal – they are primed to anticipate an experience of popular communion which may have no basis in reality. It may be no stretch of the imagination to suggest that spectators at the modern Globe pay their £5 and purchase a simulacrum of *communitas*. Lanier sums up the argument nicely:

> What the new Globe offers, the argument runs, is a historically themed postmodern simulacrum of 'authentic' Shakespeare ... that markets a depressingly familiar and deeply unprogressive sort of bardolatry, nostalgic for an imagined English past, worshipful of the civilising effects of Shakespeare, eager to see Shakespeare's plays as a repository of disinterested, unchanging truths about the

essential human condition around which audiences, now as then, have always gathered.

(2002: 165)

As Lanier himself points out, however, the myth of an authentically popular Shakespeare – while undoubtedly steeped in fiction – 'also gives expression to cultural aspirations that cannot be entirely reduced to bad faith' (2002: 166). As the rest of this chapter will show, certain arguments in favour of the collectivist theatre retain their force in the face of materialist deconstructions, which all-too-often equate mystification automatically with hegemonic interests, and assume that any gestures towards 'universal human truths' must be conservative. The materialist will naturally question the legitimacy of such mythologising, appealing as it does to the immaterial and 'metaphysical', but not all myths (problematic though they may be) serve exclusively reactionary ends. It may be that materialist accounts of collectivist nostalgia are overstating the case, throwing – to mutilate a cliché – the progressive baby out with the conservative bathwater.

There is commonly, I contend, something fundamentally progressive implied in the desire for displays of communal solidarity. Often, as Counsell argues, it 'represents a genuine disaffection with and response to a perceived deterioration in the conditions of life under mechanistic western capitalism'; he suggests that such fantasies can provide 'a rallying point for those seeking change' (1996: 177). Bruce A. McConachie writes that 'the imaginative construction of "community"' in theatre may help audiences 'to do the symbolic work of including and excluding that constitutes community', providing them with '"what if" images of potential community' (1998: 37–8). The 'potential community' imagined may of course be a reactionary one, but not *necessarily*, and certainly not exclusively, when – as in most of the cases studied here – the imagined community includes groups previously marginalised by or excluded from constructions of the 'typical' Shakespearean audience. Conventional audience behaviour is intensely class-coded, particularly in the Shakespearean theatre, and while fantasies of a 'classless' theatre are impossible to realise, they can provide a site for cultural contestation. In any case, imagining Utopias, even if they are ultimately discarded as naïve or unworkable, is surely the first step anyone takes towards acting for progressive political change.

Furthermore, in acting out the fantasy of communality, the spectator is no longer a passive consumer of a packaged spectacle, but an active participant in a social event. With the advent of 'representational' drama, as Raymond Williams once pointed out, audiences 'learned to give up the idea of intervention, participation, direct action, even as a possibility, in favour of direct, conventional and reacting forms' (1968: 185). What has become the conventional actor/audience relationship in theatre privileges performer and director as the sole creative sources of a cultural product, allowing them an unchallenged, one-way flow of communication. The passive, non-participatory audience acquiesces to an authoritarian form of theatrical communication, submitting to what is temporarily at least a 'higher' cultural authority. Participatory theatre, on the other hand, implies that culture need not be imposed from 'above', but rather created collaboratively by the assembled community; for the artist, argues Mark Weinberg, it is 'in some ways a repudiation of claims of superiority in aesthetic judgement and artistic virtuosity' (2003: 187). Of course, the power remains very much in the hands of the artists, but in foregrounding the audience's responses, the collectivist theatre does suggest that at least those responses *matter*.

The characterisation of audience response as an unquestioning acceptance of the collectivist myth is also, I think, something of a simplification. Audiences are surely aware, on the whole, of the fissure between their actual experiences and the mediatised simulations of those experiences. The films cited above provide *ideals* of Shakespearean spectatorship, but not necessarily *expectations*; they derive their emotional impact, in fact, from the perceived rarity of theatrical communion. Mediatisation may serve, paradoxically, to increase the audience's scepticism of *communitas* while simultaneously strengthening their desire for it. When Robert Shaughnessy describes 'the kind of Disneyland double consciousness simultaneously composed of amused engagement and sceptical distance' which distinguishes Globe spectatorship (2000: 3), his description is reminiscent of the way in which Susan Willis suggests the potential of historical theme parks to 'produce critical rupture with the present': 'While every encounter with the past runs the risk of recuperation,' she argues, 'those moments when we use the past to engage with the present have the power to escape nostalgia' (1991: 15).[13] It might be argued that a 'critical rupture' between actual audience experience and its nostalgic construction in the media could produce effects similar to those of the 'disjunctive anachronisms' studied in Chapter 2.

The assumption that an impression of *communitas* is always merely the result of mystification and mythologisation must also be questioned. In many cases, the advocates of a collectivist popular theatre do their cause a disservice by mythologising it. Communal response itself is no myth: it is manifest whenever an audience responds *en masse*, and though the spectators who make up the audience must respond individually and subjectively, the sum total of their responses cannot fail to characterise them as a group. From a materialist perspective, then, *communitas* might be understood as the result of a semiotic process which takes place between spectators: one sees one's own subjective response confirmed by its articulation elsewhere in the audience.[14] When such responses are experienced in public as part of a group of people, many of whom show outward signs of similar responses (particular kinds of laughter, applause, cheering, gasps, focused silence), these may, for a moment, be very powerfully felt.

In his study of *Liveness*, Philip Auslander describes his impatience with

> clichés and mystifications like 'the magic of live theatre,' the 'energy' that supposedly exists between performers and spectators in a live event, and the 'community' that live performance is often said to create among performers and spectators.
>
> (1999: 2)

Auslander's verdict typifies the attitudes of much of the materialist criticism considered in this chapter, and he indeed goes on to deconstruct the notions of 'magic', 'energy', and 'community' which he cites from a materialist perspective. I wonder, however, whether these aspects of performance *can* be understood in exclusively materialist terms. Materialist criticism of collectivist theatre, as McConachie points out, tends to

> slight spectators' general emotional response to these kinds of performances to concentrate on the purported ideological meanings read by the audience. ... Certainly ideological transaction is a part of every performance, but the mutual making of meaning is not all that occurs in theatre.
>
> (1998: 33)

Audiences *do* undergo subjective responses, and these *are* experienced as part of a group (though of course 'communal response' is always experienced individually). Materialist criticism, however, has difficulties accounting for such subjectively experienced aspects of theatre (hence the structure of this book), and it may be that commentators recourse to mythologising language in discussions of *communitas* precisely because materialist discourses provide no means for an adequate expression of it. In response to Auslander's objections, Jill Dolan writes:

> I must admit that I believe in all the things that Auslander dispar-ages, mostly because as a one-time actor, and as a director, writer, spectator, critic, and performance theorist, I've experienced them all. I've felt the magic of theatre; I've been moved by the palpable energy that performances that 'work' generate; and I've witnessed the potential of the temporary communities formed when groups of people gather to see other people labour in present, continuous time, time in which something can always go wrong.
>
> (2001: 458–9)

Dolan goes on to draw upon the anthropologist Victor Turner's account of *communitas* in order to validate her experience (2001: 473).

Turner provides a less mystical explanation than many of those deconstructed above, but not a materialist one; he analyses a type of experience which he claims '*has* to be dealt with phenomenologically' (1982: 58). *Communitas*, he argues, is neither uncritical participation nor mere myth: 'communitas preserves individual distinctiveness – it is neither regression to infancy, nor is it emotional, nor is it "merging" in fantasy' (1982: 45–6). Turner explains *communitas* in terms of ritual, which he understands according to Arnold van Gennep's three phases in a rite of passage: *separation, transition* ('limen'), and *incorporation* ('reaggregation'). Turner focuses on the transitional, 'liminal' stage of ritual, in which subjects 'pass through a period and area of ambiguity' (1982: 24). In liminality, he argues, 'people "play" with the elements of the familiar and defamiliarise them' (1982: 27), and where it is 'socially positive',

> it presents, directly or by implication, a model of human society as a homogenous, unstructured communitas, whose boundaries are ide-ally coterminous with those of the human species. When even two people believe they experience unity, all people are felt by those two, even if only for a flash, to be one.
>
> (1982: 47)

These Utopian models, argues Turner, 'can generate and store a plurality of alternative models for living' (1982: 33). He suggests that in contrast to the 'liminal' activities of tribal cultures, the 'liminoid' activities of post-industrial societies tend to be separate from the 'central economic and political processes'; the liminoid is often a commodity 'which one selects and pays for', and less often collectively experienced (1982: 54–5). However, it may be that the commodification of liminoid phenomena as 'leisure activities' allows them to serve a more progressive end than the safety-valve function of 'liminal' phenomena, as Marvin Carlson suggests:

> [B]eing more playful, more open to chance, they are also much more likely to be subversive, consciously or by accident introducing or exploring different structures that may develop into real alternatives to the status quo.
>
> (1996: 24)

For Turner, post-industrial liminality allows 'lavish scope' to artists for the generation of 'models, direct and parabolic or aesopian, that are highly critical of the *status quo* as a whole or in part' (1982: 40).

Usefully for our purposes, Turner also distinguishes between 'spontaneous' and 'ideological' *communitas*. The former he defines as 'a direct, immediate and total confrontation of human identities': 'It has something "magical" about it. Subjectively there is in it the feeling of endless power' (1982: 47–8). Essentially, it is a *momentary* phenomenon, an empowering 'flash' of optimism which 'may succumb to the dry light of the next day's disjunction, the application of singular and personal reason to the "glory" of communal understanding' (1982: 48). It should be noted, however, that Turner is not advocating 'something magical' or 'the feeling of endless power' as adequate descriptions of the phenomenon. Ideological *communitas*, he argues, is the unavoidably frustrated attempt to describe the subjective experience of spontaneous *communitas*: the experiencer, he suggests, will attempt to

> sack the inherited cultural past for models or for cultural elements drawn from the debris of past models from which he can construct a new model which will, however falteringly, replicate in words his concrete experience of spontaneous communitas.
>
> (1982: 48)

In this sense, ideological *communitas* is spontaneous *communitas* transformed into discourse. Turner's 'ideological communitas' is markedly similar to the 'simulated' *communitas* discussed earlier in this chapter. It is, I would suggest, the inevitable mythologisation of an incommunicably subjective momentary experience.

Crucial to Turner's conception of *communitas* is that it is never stable; by his own definition it is 'spontaneous', an ephemeral moment of optimism, always transitory, constantly being renegotiated. Such emotions may be experienced as part of a dialectical process of 'moving in and out' (to paraphrase Brook), and as Shomit Mitter has pointed out, *de*familiarisation is necessarily the result of a transition from what might be called 'familiarisation':

> Alienation is effective only in proportion to the emotional charge it undercuts. As Brecht once remarked to Giorgio Strehler, while the music in his theatre was designed to break the illusion, that illusion had 'first to be created, since an atmosphere could never be destroyed until it had been built up' (Brecht 1979: 102). Just as metatext depends for its existence upon text, so also estrangement depends upon identification for its effect.
>
> (1992: 49)

Group affirmation, then, is not incompatible with 'critical rupture', but rather a necessary component of it. Audiences at the Globe's *Richard III* in 2003 may have experienced a powerful moment of togetherness, for example, when, spurred on by Amanda Harris's Buckingham, they cheered for Kathryn Hunter's Richard to take the crown; the moment was described by John Gross as 'an example of Globe audience participation at its best' (*Sunday Telegraph*, 15 June 2003). But as the play progressed, it became increasingly apparent that they had become complicit in a villain's rise to success. This created a particularly interesting tension as the full implications of Richard's kingship were played out.

Practice at the Globe might point towards a more complex interplay between spontaneous and ideological *communitas* than Turner's analysis suggests. As we have seen, audience behaviour may be shaped by mediatised expectations (ideological *communitas*?) and encouraged, as in the example above, from the stage. But as Pauline Kiernan's note on audience responses at the first previews at the Globe implies, it may ultimately be impossible to separate the myth from genuinely felt subjective experience:

> Is this just contrived 'joining in'? Sometimes it seems manufactured, sometimes it's natural. Perhaps playgoing in an open playhouse simply provokes a 'natural' desire to 'manufacture' this response? And if it is contrived, does it matter? At moments when

the audience is emotionally moved and goes quiet, its participation is as strong.

(1999: 112)

What is for one audience member a 'manufactured' response may seem 'natural' for another; a manufactured response, Kiernan suggests, may in fact *lead* to a 'natural' response later on. The binary becomes unsustainable. *Communitas*, or the simulacrum of *communitas*? If both are experienced individually, is there, in the end, any difference?

Personal Narrative 6
Alternative Endings

Ending one: *Titus Andronicus,* Royal Shakespeare Theatre, 16 June 2006

Yukio Ninagawa's Japanese-language production of *Titus Andronicus* is just ending. Epic, spectacular, it has played out over three-and-a-half hours on a massive white stage with a cast of 30. The acting has been formal, physical, the delivery at times verging on the operatic; Shakespeare's text has scrolled across the sides of the stage on surtitles, though by this point I have given up on following it (I know the play well enough, after all), and focused my attention upon the captivatingly beautiful stage pictures. The play's violent climax, making use of symbolic scarlet ribbons for the blood, has been almost balletic. The 'flawed redeemer' Lucius has just left the stage, ready to 'heal Rome's harms and wipe away her woe' (5.3.147), but not before passing his vicious final sentences upon Aaron and the corpse of Tamora: the dead queen is to be thrown 'to beasts and birds to prey' (5.3.197), and her lover set 'breast-deep in earth' and starved (5.3.178).

The stage empties. Only Lucius' son, Young Lucius, and the soon-to-be-orphaned baby of Tamora and Aaron remain. This is the baby that only minutes ago, Lucius threatened to hang; he has relented, of course, promising Aaron that he will 'nurse and bring him up' (5.1.84), but he showed at best a disinterest in its well-being as he left the stage.

Young Lucius holds the child of his family's enemies in his arms.

He screams.

He screams until he runs out of breath, and then he screams again.

He screams again.

He screams again.

A numbing, prickling sensation spreads through the roots of my hair. The image embodies something horrifying about the results of comparable cycles of violence which beam onto our television screens from all over the world today.

The play concludes, and as the cast return to take their bows, the auditorium erupts with applause. As those around me stand to register their approval, I find myself on my feet too.

Ending two: *A Midsummer Night's Dream*, Swan Theatre, 17 June 2006

One day later and now Tim Supple's 'Indian' production of *A Midsummer Night's Dream* is also finishing. Like last night's performance of *Titus*, it is one of the international contributions to the RSC's *Complete Works* festival (the programme tells me that its script includes 'English, Tamil, Malayalam, Sinhalese, Hindi, Bengali, Marathi, and even a smattering of Sanskrit'); like *Titus*, it too has been ravishingly spectacular. But there the similarities end.

Supple's ritualistic *Dream* fills the more intimate space of the Swan with colour and activity: the stage is dusted with dry red earth, a bamboo scaffold lined with ripped paper stretches across the back wall, and long, richly-coloured silks, upon which the performers have spent the evening performing graceful aerial stunts, hang from the ceiling. The acting has been athletic, sensuous, earthy. Since Ajay Kumar's Puck ran his hands across a large, round stone in the centre of the theatre to produce an ethereal hum, the piece has registered itself very strongly as the domain of the 'magical'.

We reach the end. Oberon, Titania and their acrobatic fairies return to 'sing and bless this place' (5.2.30); Puck appeals directly to the audience to 'give me your hands, if we be friends' (Epilogue, 15). The entire company begins to sing in harmony, dancing slowly and rhythmically, stamping their feet in unison, smiling as they do so. They are unified in both voice and movement.

Once again, members of the audience are on their feet: this time, though, they are clapping and moving in time with the music. The song ends, and the theatre reverberates with the noise of applause and cheering; the woman next to me shouts 'Bravo!' It is, it seems, a joyful celebration of a culture-transcending optimism.

The production has been beautiful and uplifting. I feel like a killjoy, but for some reason, I can't bring myself to participate in a standing ovation for a second consecutive night.

6
Shakespeare, Space, and the 'Popular'

> I can take any empty space and call it a bare stage. A man walks across this empty space whilst someone else is watching him, and this is all that is needed for an act of theatre to be engaged.
>
> (Brook 1990: 11)

'Empty' spaces

Peter Brook's seminal pronouncement in the opening lines of *The Empty Space* has provoked reams of commentary, to which I do not wish to add at any great length. I quote it simply to point out that it reflects a general assumption about theatre space which is more often than not taken for granted: that prior to its inhabitation by a performance, a theatre space *may* be empty. Try putting Brook's exercise into practice, however, and one finds it almost impossible to get started; the spaces available are not empty, but littered rehearsal spaces or decorated halls, busy classrooms, or inhospitable fields. Brook's 'empty' space is, crucially, an imaginary one.

The opening sentence of Henri Lefebvre's *The Production of Space* also refers to emptiness, but in a very different context:

> Not many years ago, the word 'space' had a strictly geometrical meaning: the idea it evoked was simply that of an empty area.
>
> (1991: 1)

Lefebvre goes on to challenge such a conception of space. His central thesis is that space can never be empty, or passive, but rather that '(social) space is a (social) product'; space, he argues, 'serves as a tool of thought and of action', and that 'in addition to being a means of production it is also a means of control, and hence of domination'. However, the very fact that space is always *in process* means that 'the social and political (state) forces which engendered this space now seek, but fail, to master it completely' (1991: 26). Thus, in Lefebvre's model, space is divided into two broad categories: 'dominated space', which is controlled and imposes hegemonic ideology; and 'appropriated space', which is adapted by a specific class or group of people according to their needs (1991: 165).

Applied to theatre (and also, in this chapter, specifically to venues for Shakespeare), Lefebvre's model forces one to consider the interests served by performance space. Concerned as we are with the 'popular', we begin to see how not only architecture but also factors such as ticket price, geographical location, and publicity define a theatrical event just as much as what goes on onstage, and often in the service of the perpetuation of an artistic or social elite among audiences. As John McGrath puts it in *A Good Night Out*, theatre is a 'very complex social event', and when artists see such issues as 'someone else's problem', they relinquish 'a great deal of the meaning of the event socially and politically' (1996: 4–5).

Theatre space can be 'dominated' before its audience has even set foot inside it. Expectations of a theatre building can be determined by representations in the media (as we saw in the previous chapter), by previous experiences (Iain Mackintosh describes childhood memories of 'cramped and badly run bars, rude box office staff imprisoned behind wrought-iron guichets and uncomfortably close-spaced seats'; 1993: 79), or simply by general patterns of social behaviour. Director Pete Talbot explained to me some of the difficulties encountered by his company during regional open-air touring:

> [W]e get frustrated sometimes that we get so many kinds of older, bourgeois, middle-class people, and we would prefer to be able to have a wider audience. The problem isn't because our theatre is unsuitable for people who don't normally go to the theatre; it's because they don't think of coming.
>
> (Talbot 2003)

The theatrical meaning-making process is also, of course, defined in part by location. For example, a theatre such as the Watermill (buried in the countryside near Newbury) which is accessible only by car automatically excludes potential audience members without access to private transport; a production on a university campus will almost una-voidably define itself as part of the world of education (Gay McAuley cites accounts of audience members who see such locations as part of an uncomfortably 'alien world'; 2000: 47). Performances on village greens or in village halls will generally have a strongly 'local' flavour, even if the production is by a professional touring company such as McGrath's 7:84, while theatres in regional towns will often be considered by local residents to reflect the wider cultural status of the community of which they are a part, and as such a symbol of civic pride (or otherwise).

Specific towns, of course, also have specific connotations, and when it comes to Shakespeare, Stratford-upon-Avon enjoys a distinguished posi-tion as the poet's 'home town'. Tours of Shakespeare's Birthplace are sold to the same tourists buying tickets for productions at the RST, lending the theatre a unique claim towards being the metaphorical 'home' of his works, and the reverent, almost religious veneration exhibited towards Shakespeare (the man) at the Birthplace during the day is extended to the theatrical experience at the playhouse in the evening. As the RSC's former Associate Director Michael Attenborough has put it:

> There's a sense of occasion about theatre-going in Stratford. People have made a pilgrimage to see the work.
>
> (Mulryne & Shewring 1995: 91)

The reconstruction of Shakespeare's Globe in Shakespeare's 'other' home, Southwark, has led to a new challenge to Stratford's primacy as the 'original' Shakespearean venue, one with at once a greater claim to legitimacy as the 'actual' home of the plays themselves, and a much lesser one as a 'mere reconstruction'. Further discussion of the new Globe will follow later.

Within a city, the area in which a theatre venue is situated is crucial to its connotative meanings. In London, for example, the Soho Theatre is imbued with all the subversive, underground energies of its surrounding area, while the Arcola's situation in the impoverished but increasingly arts-oriented Hackney marks it out as a distinctly 'fringe' venue; the many pub-theatres on London's fringe scene exhibit their counter-cultural status just as firmly. A West End location, of course, advertises a production as being at the very centre of the theatrical map (both liter-ally and metaphorically), but in a very different manner from the no less 'central' South Bank, where productions at the National Theatre, the Old and Young Vics, and the Globe seem to define themselves in primarily cultural rather than commercial terms. Theatre locations are clearly suffused with ideological implications. When Richard Eyre describes the Royal Court Theatre as 'perfectly placed between the (now) ersatz Bohemianism of Chelsea, and the wealthy austerity of Belgravia' (1993: 141), he plainly betrays his own notion of theatre's 'perfect' location within culture in general as situated between pretentious nonconformity and bourgeois self-importance – an ideal which anyone interested in popular theatre might wish to contest. However, one must be wary of jumping to the easy assumption that location within a working-class district will attract a more 'popular' audience in the anthropological sense of the word: Joan Littlewood's experiments in Stratford East, among many other attempts to bring theatre to working class audiences, have shown that this is not always the case (see McAuley 2000: 45). Pete Talbot recounts his experience of bringing classical theatre to a council estate in Tadley, Hampshire, as part of Basingstoke and Deane Borough Council's festival: 'nobody turned up. The only people who came were people who were not from the area' (Talbot 2003).

Previous chapters have discussed the way in which an emphasis upon the dual nature of theatre is frequently constructed as the defining feature of a popular performance (whether to unify the audience in a 'shared experience', or to instil in them a dialogic attitude towards the performance). One of the primary concerns of this chapter, then, is the role played in such processes by theatre space. Certain principles are obvious. Large spaces relegate lower-paying customers to seats a very long way from the stage, which is inimical to theatrical 'communion'; sources from the ancient Sanskrit holy text the *Natya Shastra*, through Laurence Oliver, to today's leading theatre architects, agree that the maximum distance between the actor and the last row of the audience should be no more than 16 to 21 metres.[1] Irish playwright Dion Boucicault once stated categorically that 'the popular places of public entertainment have always been medium-sized or small' (1889, quoted Mackintosh 1993: 125); more recently Oliver Double has described the advantages of low ceilings for the creation and containment of laughter (1997: 161), a sentiment which has been echoed by William Cook (2001: 110).

The way in which the theatre space is organised determines to a very great extent what is 'shared' and what is compartmentalised. First, there is the division between what McAuley calls 'audience space' and 'practitioner space' (2000: 25–6). 'Backstage' is usually hidden away from the audience, obscuring the process which creates the performance and, as John Russell Brown suggests, implying 'a distance, or even a gulf' between actor and audience (1999: 159). Stage doors are often situated a long way from the audience entrances, leading inconspicuously into corridors which from the outside may not even be an obvious part of the theatre building (there are, of course, exceptions – actors leave the West Yorkshire Playhouse alongside audiences, and at La Cartoucherie, the Paris home of le Théâtre du Soleil, even the dressing rooms are visible to playgoers). The conventional theatre building also divides the audience itself. Most proscenium-arch theatres grade their seats by price, charging higher amounts for the tickets with the best and/or closest views of the stage, and less for the seats at the back, at the sides, behind pillars, or in the higher galleries. The cheapest tickets tend to be those with the most restricted views (though some theatres offer special deals for on-the-day purchases). Such a system – typical of the commercial theatre in particular – clearly privileges wealthier playgoers in terms of comfort and visibility, but crucially it also separates them from the less well-off; in some theatres, the cheaper galleries even have their own entrances via side-doors from the street outside, giving those customers who have paid less the distinct feeling that their ticket is very much 'second-class'. Indeed, this division is surely comparable to the old Victorian townhouse with its separate entrance for 'tradesmen'. Such acute social division is reflected in the term 'Dress Circle', where in bygone ages evening dress was expected to be worn.

Even in modern auditoriums, division is rife. At the Barbican or the Lyttleton, it is possible to sit in the stalls without even noticing the presence of circles or galleries above and behind you; the long rows of seats in almost-straight lines mean that members of an audience are focused very much on the stage and not each other. Swedish director and designer Per Edström identifies two key patterns in theatre formations: on the one hand, the picture-frame stage, bordered along one side by straight rows (what he calls the 'monologue' form, since 'people gather in front of a picture or a man giving a monologue'); on the other hand, the open stage, surrounded by audience members on all sides (what he calls the 'action' form, since 'people gather on all sides around men acting with swords or words'; 1990:13). He makes his view of the ideology behind such use of space very clear:

> A theatre group that wants to convince an audience about their ideas without giving them the opportunity to debate these ideas avoids the circle of debate form, and chooses a monologue and picture theatre form in order to placate them and indoctrinate them with words. A theatre group that wants to share their experiences of life with an audience chooses to gather the audience around themselves so that the theatre performance can be a form of debate, containing giving

and taking, which will give the theatre group new experiences to share with a new audience.

(Edström 1990: 17)

His analysis, of course, recalls Bakhtin's distinction between 'monologic' and 'dialogic' forms. If we are to work towards a definition of a 'popular' dramatic space in an anthropological sense, then, we must locate a space in which a sense of ownership is shared by both actors and audience (rather than, for example, by theatre management companies, producers, or sponsors). Based on the factors considered so far, it seems that what might be called 'democratic' configurations (circles and thrust stages), intimacy, easy access, an absence of division, and a sense of 'appropriation' will be the key factors in such a space. These issues are explored further in the sections on 'Collectivist space' and 'Appropriated space' below.

Elizabethan theatre space

Since we are concerned with spaces for Shakespeare, it seems sensible to look for a moment at the kinds of theatre space for which the plays were originally written. This will, however, be a necessarily brief overview: so much has been written on the architecture of the Elizabethan playhouses that it is hardly worth retreading such well worn ground here in any great detail.[2]

The basic facts, as we know them, are these. Playhouses were divided into two main categories: the outdoor public theatres and the indoor private ones. The private playhouses, developed from the Tudor domestic hall and at first the domain of companies of child actors, were not used by adult companies until 1608. The public playhouses (for which most, but not all, of Shakespeare's plays were written) were generally circular or polygonal in form, open-roofed, and characterised by a yard or pit surrounded by galleries on two or three levels. A thrust platform stage (usually rectangular, but lozenge-shaped in some cases) jutted out into the yard, surrounded on three sides by standing playgoers and on all sides by galleries. Behind it was the 'tiring house', and above it, an elaborately decorated roof. Thomas Platter's much-quoted contemporary account gives a good idea of the social organisation of the spaces:

> The places are so built, that they play on a raised platform, and every one can well see it all. There are, however, separate galleries and there one stands more comfortably and moreover can sit, but one

pays more for it. Thus anyone who remains on the level standing pays only one English penny: but if he wants to sit, he is let in at a further door, and there he gives another penny. If he desires to sit on a cushion in the most comfortable place of all, where he not only sees everything well, but can also be seen, then he gives yet another English penny at another door.

(Chambers 1923: II, 365)

The physical layout of the Elizabethan amphitheatre implies a use of space which is very different from the nineteenth-century theatre conventions we take for granted today. As Pauline Kiernan points out, with its vertical configuration of galleries and its throng of 'groundlings' below the stage, the Elizabethan amphitheatre forms a sphere with the actor at its centre (Kiernan 1999: 18). Occasional Globe workshop director Ildiko Solti has described this phenomenon as a 'kinesphere' in which the audience's focus upon the actor is intensified by the visible lines of attention pointing towards him from all angles (Solti 2004). Certainly, in daylight, other members of the audience must have formed a highly visible backdrop to the actor, wherever a playgoer was positioned. Thomas Middleton's reference to the audience in *The Roaring Girl* (1611) seems to testify to this effect:

Within one square a thousand heads are laid,
So close that all of heads the room seems made.

(1.1.140–1)

The effect of this, of course, was that as Brecht put it, 'people were supposed to use their imaginations' (1965: 58–9). Brown quotes the diary of Simon Forman, who saw a performance of *Macbeth* at the Globe on 20 April 1611:

There was to be observed, first, how Macbeth and Banquo, two noblemen of Scotland, riding through a wood, there stood before them three women, fairies or nymphs, and saluted Macbeth.

'Of course,' says Brown, 'no "wood" had been placed on the stage through which the actors could ride and in this scene no one makes even a passing verbal reference to trees; nor were any horses on stage or referred to in the text' (1999: 179). The passage, he suggests, shows the extent to which illusion was created by the imaginative willingness of the audience, and certainly the plays, the acting style, and the buildings themselves actively *required* imaginative collaboration from the audience. Says Kiernan: 'The physical conditions of the Globe structure work to reinforce the fiction's invitation to the audience to "piece out our imperfections with your thoughts"' (1999: 27).

The ideologies implicit in Elizabethan theatre space are fascinating, and all-too-easily simplified into a nostalgic fantasy of what theatre critic John Peter has described as 'an essential social unity' (*Sunday Times*, 15 June 1997). As we saw in the last chapter, this was never the case: Platter's account, in fact, shows clearly how the playhouse was segregated by wealth, and contemporary playtexts are full of disparaging references to groundlings (see Chambers 1923: II, 522–55). However, the compartmentalisation of the playhouse was the direct reverse of today's theatres. Wealthy playgoers sat in the highest galleries, while the lower classes were the closest to the action.

Naturally, this made for some unique interplay between 'high' and 'low', and this is implied in the texts themselves. 'Low' characters like rustics and clowns have asides and sequences of direct address which need to be addressed to nearby audience members, while 'high' characters' lines seem naturally to gravitate towards the upper galleries – particularly when on their first appearances they refer to the sun, moon, or stars. Three short examples will illustrate my point:

THESEUS.   Now, fair Hippolyta, our nuptial hour
　　　Draws on apace. Four happy days bring in
　　　Another moon – but O, methinks how slow
　　　This old moon wanes!

　　　　　(*A Midsummer Night's Dream*, 1.1.1–4)

BEDFORD.   Hung be the heavens with black! Yield, day, to night!
　　　Comets, importing change of times and states,
　　　Brandish your crystal tresses in the sky,
　　　And with them scourge the bad revolting stars
　　　That have consented unto Henry's death.

　　　　　　(*1 Henry VI*, 1.1.1–5)

POLIXENES.   Nine changes of the wat'ry star hath been
　　　The shepherd's note since we have left our throne
　　　Without a burthen.

　　　　　　(*The Winter's Tale*, 1.2.1–3)

Some of the more spectacular speeches, of course, encompass the entire spectrum within a short space of time. The prologue to *Henry V* seems to me to be an example: the lines 'O for a muse of fire, that would ascend / The brightest heaven of invention' surely require a delivery which projects the words upwards (towards 'heaven', perhaps), while the question 'Can this cock-pit hold / The vasty fields of France?', with its implied reference to the yard, would presumably be directed down towards the groundlings (where as we have seen, at the new Globe, it was frequently met with an audible 'yes'; Kiernan 1999: 27).[3]

Central to this interplay is the symbolic function of the playhouse as a microcosm of the Renaissance model of the universe. Lefebvre summarises the representation of the cosmos suggested by medieval figures such as Dante and Thomas Aquinas as follows:

> A fixed sphere within a finite space, diametrically bisected by the surface of the Earth; below this surface, the fires of Hell; above it, in the upper half of the sphere, the Firmament – a cupola bearing the fixed stars and the circling planets – and a space criss-crossed by divine messages and messengers and filled by the radiant Glory of the Trinity.
>
> (1991: 45)

Its correspondence to the structure of the public Elizabethan playhouse hardly needs further elaboration. In identifying the decorated roof and upper galleries with heaven, the platform stage with the earth (a metaphor drawn frequently within the texts), and the trapdoor and the pit with hell, the 'popular' elements of performance were associated with the earthly and the sinful (an association inherited from the medieval mystery cycles) and the more 'virtuous' characters separated from their fellow human beings.

This leads us on to one final point which is vital to an understanding of the plays' original spatial functions: the interplay of *locus* and *platea*. As we saw in Chapter 1, Weimann suggests that while Shakespeare's stage was not a direct descendant of medieval *platea* space, 'it did take on and expand some of the *platea's* basic functions' (1987: 80); indeed, as we have seen, Weimann argues that the interplay between *locus* and *platea* was fundamental to Shakeapeare's dramaturgy, and moreover to the Elizabethan theatre in general. He elaborates:

> Such an interplay accommodates action that is both nonillusionistic and near the audience (corresponding to the 'place') and a more

illusionistic, localised action sometimes taking place in a discovery space, scaffold, tent, or other *loci* (corresponding to the medieval *sedes*). Between these extremes lay the broad and very flexible range of dramatic possibilities so skilfully developed by the popular Renaissance dramatist.

(1987: 212)

This broad range of possibilities produces the striking contrasts and moments of self-reflexivity which characterise many of Shakespeare's plays. John Russell Brown has picked up from Weimann's ideas in several of his studies, listing in *New Sites For Shakespeare* some of the moments in *Macbeth* and *King Lear* at which the interplay is at its most effective (1999: 99–100), and arguing in *Shakespeare's Plays in Performance* that an 'open dramatic composition' in which the actor has 'no background to his figure except the anonymous audience at the other side of the theatre' might suggest 'continuations beyond the bounds of the stage' (1993: 151–2).

Conventional space

The 'de-popularisation' of theatre space in Britain arguably began before Shakespeare wrote his first play, with the construction of dedicated theatre buildings – the first being James Burbage's 'Theatre' in 1576. Later in the same year, the first indoor private playhouse was opened at the Blackfriars Dominican monastery, and as the sixteenth century turned to the seventeenth, an increasing number of such playhouses began to appear. This all came to a halt, of course, with the closure of the theatres by order of Parliament in 1642.

When Charles II was restored to the throne in 1660, the two theatres opened by royal patent under the management of Davenant and Killigrew had a new feature: where, in the Elizabethan playhouse, the 'tiring house' would have stood was now an open, scenic space, designed for the accommodation of Italian-style perspective scenery (though acting still took place primarily on the old-fashioned forestage at this point). Another marked difference was the increasingly (though by no means exclusively) aristocratic composition of the audiences entailed by the king's legitimisation of only two theatre companies; as such, Shakespeare was 'officially' performed for only a very select audience.

'Popular' Shakespeare was not entirely eradicated, however – merely illegitimatised. Shortened versions of his plays which had been performed throughout Cromwell's interregnum continued to be performed in 'illegitimate' venues such as fairgrounds and taverns during the Restoration and ensuing years. These shows, known as 'drolls', lifted popular comic characters such as Bottom, Falstaff and the Gravediggers from *Hamlet* from their source plays and re-centralised them as principal characters in their own short farces. The years which followed saw an increasingly 'illegitimate' popular Shakespeare emerge. In his history of the open stage, Southern states that

> [a]fter the large forestage of the Restoration theatre we had admittedly a gradual shrinking of the forward acting-area through the Georgian period – but something else arose to preserve the idea instead, namely the separate arena theatres such as Astley's and the Royal Circus, where the forestage was virtually increased and, indeed, became a full arena occupying the pit, with the stage a separate entity beyond.
>
> (1953: 37)

Interestingly, Southern's analysis depoliticises this historical shift. What he fails to note is that Astley's Amphitheatre and the Royal Circus were emphatically lowbrow, popular, non-mimetic spaces; when Shakespeare was presented at Astley's, it was in hybridised form, with William Cooke's spectacular horseback productions of *Macbeth* and *Richard III* (Joseph 1968: 29). Other bastardised versions of Shakespeare, featuring song, music, dance, stunts, and clowning, were performed in unlicensed theatres (often converted inns and taverns) all around Britain throughout the first half of the nineteenth century.[4]

Shakespearean performance within the 'legitimate' theatre was an entirely different matter. The eighteenth century saw the development of the proscenium arch, and gradually the forestage began to shrink into the scenic space. By the nineteenth century, an increased demand for realism meant that virtually all the acting now took place behind the proscenium arch, and as such the forestage was removed entirely to make way for additional rows of seats (see Southern 1953: 51–2). By the mid-nineteenth century, the theatre was no longer a participatory space: the auditorium was dominated by passive (and distinctly class-coded) behaviour. It is the style of 'fourth-wall' naturalism inherited from the nineteenth-century theatre which continues to dominate Shakespearean performance in conventional proscenium-arch theatre spaces today, with actors often playing whole sequences 'as if there wasn't an audience' (Brecht 1965: 51).

Objections to this are twofold. From the 'collectivist' perspective, as Per Edström argues, naturalism's 'utter lack of contact and absurd conventions of pretence destroy far too much of the active and living contact between actors that puts life into the art of theatre' (1990: 93). Referring to the speech in which Leontes comments on the many men who are 'even at this present, / Now, while I speak this' (*The Winter's Tale*, 1.2.193–4), Southern argues that the proscenium arch is fundamentally unsuited to Shakespeare:

> There are many and many such passages in plays written for open stages which lose more than half their impact when spoken as the pale reflections of a thinker in the 'removed room' of a picture-frame stage. They do not then touch *us*; we can escape them.
>
> (1953: 45)[5]

From the 'critical' perspective, as Graham Holderness argues, such theatre allows ideology 'a free and unhampered passage to the spectator':

> As the stage becomes more illusionistic, it permits less space for the collaborative creation of meaning natural to the Elizabethan theatre. As the audience is further removed from the action, it becomes a passive consumer of a fixed ideology rather than an active constituency intensely involved in a complex process of reciprocal communication in which ideologies can be interrogated, contradictions made visible, conventions subverted and orthodoxies exposed.
>
> (2001: 46–7)

It might be argued, in fact, that the retreat of the forestage and advance of the proscenium arch into the audience space was an indirect result of the rise of capitalism. As *platea* space became extraneous to theatrical performance, both the collective act of imagination and the dynamic public clash of discourses were gradually replaced by the consumable imaginative product.[6]

While it is indeed rare for a Shakespearean production in a proscenium-arch theatre to achieve any sense of continuity with the popular tradition of *platea* acting, there are exceptions. Often, however, having transferred from other spaces, such productions do not sit particularly well in the new building. Vesturport's *Romeo and Juliet* moved from the Young Vic, where it was performed in traverse, to the West End's Playhouse Theatre, and in this new venue its lit auditorium and outrageous audience interaction – which had seemed vibrant and exciting in its original location – became a little forced and uncomfortable, somehow violating the unwritten contract between audience and performer implied by the more conventional space. Similarly, Shakespearean productions by Edward Hall's company Propeller, conceived for the intimate galleried space at the Watermill, Newbury, often seem ill at ease when they tour to regional Victorian theatres. Hall frequently has performers making exits and entrances through the auditorium, but the status of the space through which they are passing often seems to be not so much *platea* as ineffectively defined *locus*. The most successful subversion of conventional space I have seen this company perform was, in fact, not in the auditorium but in the theatre foyer: during the interval for their 2003 production of *A Midsummer Night's Dream* at the Comedy Theatre, the entire cast invaded this 'audience space' to provide a close harmony, *a capella* rendition of a series of 1960s pop songs. The sense of 'shared experience' engendered by this act was far more pronounced than any blurring of the stage boundaries they undertook during the performance itself.[7]

The most successful subversions of the 'dominated space' of proscenium-arch theatres tend to be when the implicit rules of the space are not so much transgressed as utterly disregarded. Max Stafford-Clark's touring *Macbeth* in 2004 did 'site-specific' promenade performances in spaces ranging from abandoned factories to stately homes, and when it came to conventional theatre spaces it treated them no differently. At the Oxford Playhouse, spectators did not take seats in the auditorium, but were led in a walkabout performance from the theatre's workshop to the stage itself. Stafford-Clark's rationalisation of this was, as we might expect, concerned with reacting against the lazy consumerism inherent in conventional theatre space:

> It's possible to watch, say, *King Lear*, see Gloucester's eyes cut out and then go to the bar, order a gin and tonic and say 'how moving, how very affecting,' but you're not necessarily involved. A promenade or site-specific production dislocates the usual relationship between actors and audience and moves both parties out of the comfort zone.
>
> (Out of Joint 2005)

Another production to (almost literally) explode a conventional theatre space was Jonathan Kent's *Tempest* at the Almeida, just prior to the theatre's closing for refurbishment in 2000. Audiences entered to find the stalls flooded and littered with rubble, the roof partially open to the elements, and the Front of House staff, somewhat unsettlingly, wearing hard hats. The partial destruction of the theatre building lent a particularly metatheatrical edge to the production. In Brechtian terms, perhaps, we had been 'estranged' from our usual relationship with the space.

'Collectivist' space

If the nature of audience engagement is determined in part by the theatre space itself, then, it seems the picture-frame stage is in most cases unsuitable for 'popular' Shakespeare. Popular engagement of either the collectivist or the critical variety must be, and indeed has been, scripted far more successfully in other types of venue. I should like, then, first of all, to look at what might be described as 'collectivist' space.

A renewed stress on the importance of creating a sense of fellowship among audiences was felt quite early on in the twentieth century. In his book *The Exemplary Theatre* (1922), Harley Granville-Barker protested against

> the pattern of theatre which ranks the seats in long straight rows, admirable for a view of the stage but – and this consideration has been characteristically neglected in a time when everything has been forgotten about the drama except that it is a paid entertainment – nullifying any friendly relation in the audience to each other.
>
> (1922: 215–6)

He concludes that 'a double question is involved: the physical focussing of attention, and the relative importance of one's own concentration upon the play and of being in touch with one's neighbours' (1922: 217). Granville-Barker was perhaps led in this sentiment by his early work with William Poel's Elizabethan Stage Society (1894–1905). Poel was a passionate proponent of the use of thrust staging for Elizabethan drama, and began experimenting with it as early as 1881 – a practice he continued throughout his long career, visiting halls and courtyards with non-scenic performances acted in shared light upon very basic platform stages. Poel's work was frequently derided and often passed without much recognition at all; it did, however, win him such supporters as George Bernard Shaw, who wrote that

> [t]he more I see of these performances by the English Stage Society, the more I am convinced that their method of presenting an Elizabethan play is not only the right method for that particular sort

of play but that any play performed on a platform amidst the audience gets closer home to its hearers than when it is presented as a picture framed by a proscenium.

(Mackintosh 1993: 52)

The end result of Poel's experiments was, as Stanley Wells put it in a programme note for the 2005 William Poel Festival, 'a vastly increased understanding of early stage craft' and the inspiration for a minor revolution in theatre building later in the twentieth century (Wells 2005).

Poel's ideas were picked up by director Tyrone Guthrie, who was similarly convinced of the merits of open staging – particularly as far as Shakespeare was concerned. Guthrie's autobiography *A Life in the Theatre* describes his experience improvising a new indoor staging for his 1936 production of *Hamlet*, when sudden torrential rain prevented the performance from taking place in the courtyard of Kronberg Castle, Elsinore. Arranging chairs in a circle around a central acting space in the ballroom of a nearby hotel ('the phrase hadn't yet been invented, but this would be theatre in the round'; 1961: 170), Guthrie found that what followed strengthened him in his conviction that 'for Shakespeare, the proscenium stage is unsatisfactory':

At its best moments that performance in the ballroom related the audience to a Shakespeare play in a different, and, I thought, more logical, satisfactory and effective way than can ever be achieved in a theatre of what is still regarded as orthodox design.

(Guthrie 1961: 172)

Guthrie went on first to convert the Assembly Hall, Edinburgh into a thrust-stage auditorium, and then to play a major role in the design and construction of two further thrust-stage theatres: the Festival Theatre at Stratford, Ontario, and the Tyrone Guthrie Theatre at Minneapolis. His work was also a profound influence on the developments of the Chichester Festival Theatre (1963) and the Crucible in Sheffield (1971).

A third important figure in the championing of open staging – in this case, not only the thrust stage, but also in-the-round – was Stephen Joseph. Not as well known as Poel or Guthrie, but equally pioneering in his own way, Joseph opened Britain's first modern in-the-round theatre in the upper floor of a Scarborough library in 1955.[8] In his own words, Joseph's priority in designing the theatre was that 'the presence of the actor is more strongly felt and the contribution of the audience is increased' (1968: 19). Joseph, too, saw Shakespeare as particularly suited to the open stage

> partly because of the language with its magnificent images, its soliloquies and purple patches; partly because of the rapidity with which the action is pursued, the swift changes of scene, the vigorous interplay of dynamic characters; partly because of the key in which these plays are written, by no means realistic but concentrated, enlarged and splendid.
>
> (1968: 55)

Fundamentally, when audience members can see not only the actor, but also one another, a different kind of imaginative activity must be undertaken: one might recall the model of 'shared experience' theatre discussed in Chapter 5. 'A theatre,' argues Stephen Joseph in his book on *New Theatre Forms*, 'is a building designed to house actors and audience so that the latter can see (and hear) the former' (1968: 119). Given the issues discussed above, one might have expected Joseph to define his theatres very differently; the theatre buildings we are concerned with here were designed to house actors and audience so that both parties can see and hear *each other*.

As we have seen, Per Edström identifies the picture-frame stage with autocratic monologues, and the three- or four-sided stage with democratic debate. While perhaps somewhat simplistic, Edström's argument is supported by the architecture of senate buildings both modern and ancient: arranging a space so that every person in it can see almost every other person is much more empowering for the majority than straight rows all pointing in the same direction. Bakhtin quotes Goethe's description in *Italian Journey* of the self-awareness brought to the crowd inside Verona's amphitheatre:

> [T]his many-headed, many-minded, fickle, blundering monster suddenly sees itself united as one noble assembly, welded into one mass, a single body animated by a single spirit.
>
> (1984: 255)

In what he calls a 'carnivalesque organisation of the crowd', Bakhtin argues that

> the people become aware of their sensual, material bodily unity and community. ... In the world of carnival, the awareness of the people's

immortality is combined with the realisation that established authority and truth are relative.

(1984: 255–6)

In such cases, the 'festive' organisation of theatre space becomes 'popular'
in a very political sense of the word.

The essential geometrical feature of 'collectivist' theatrical space, then,
might seem to be a circular formation. Certainly Edström shows, with geometrical diagrams, how all his theatre formations are formed from the basis
of overlapping circles (1990: 187–95), and circular patterns are central to
the work of Peter Brook and to the layout of the traditional circus. Indeed,
venues from the Royal Exchange, Manchester to the new Shakespeare's
Globe have presented distinctly 'collectivist' Shakespearean productions
in the centre of a ring of spectators, and Richard Olivier, director of *Henry
V* during the new Globe's opening season, has suggested that '[a] circle is
inclusive by its nature. No one is left out' (Kiernan 1999: 141).

But the circular formation does have its drawbacks. While a long,
sloping, circular or semi-circular auditorium (in the manner of a Greek
amphitheatre) might seem at first glance to be the most democratic
configuration of audience space, there must necessarily in larger auditoriums be a significant distance separating the front row from the rows at
the back. Furthermore, Mulryne and Shewring have suggested that

> [t]he allusion to Greek amphitheatres in the design of important
> theatre buildings in recent decades ... can be read as a wish-fulfilling
> return to the community-based values for which the Greek theatres
> are said to stand.

(1995: 11)

They argue, however, that this provides a representation of a fundamentally *false* community, since

> [t]he fragmentation of our values, as well as the physical conditions of
> the amphitheatre houses themselves, and the training and skills of our
> actors in an age of cinema and television, have ensured that only excep-
> tionally have these amphitheatres, outside opera and musical, achieved
> a sense of common experience and purpose among audiences.

(1995: 11–12)

A further argument against the pure circle derives from its drawbacks
in purely dramaturgical terms; designer William Dudley points out that

from an actor's point of view, 'if you are not at the point of command in a circle, that is dead centre, the actor can feel dynamically weak' (Mulryne & Shewring 1995: 98).

The key feature which the more successful in-the-round theatres, such as the Royal Exchange or the Globe, share with many non-circular 'collectivist' spaces is the presence of galleries. Mulryne and Shewring argue that the courtyard form has had more success than the amphitheatre 'in evoking and confirming a sense of audience engagement or "community"', citing the Cottesloe, the Tricycle in Kilburn, the RSC's Other Place and the Courtyard Theatre in Leeds (1995: 12). Certainly the contributions to their collection *Making Space for Theatre* by various luminaries of contemporary British theatre testify to their argument. Architect Michael Reardon recalls that he based the design of Stratford's Swan Theatre upon the architecture of galleried churches, because he felt that such buildings 'had the power to draw the audience together in a way that modern theatres do not' (1995: 25); scholar and critic Robert Hewison, meanwhile, describes the Young Vic's gallery as 'creating a genuine sense of theatre-as-forum' (and he notes that 'Shakespeare has usually done well here'; 1995: 54). Throughout the book, contributors give the impression that theatre directors and designers in general seem to have great affection for courtyard theatres (especially, it seems, the Cottesloe, the Tricycle, and the Swan) and an aversion to the Lyttleton in particular (a theatre with a 'terribly wintry feel', according to Cheek By Jowl director Declan Donnellan; 106). Globe associate director Tim Carroll appeared to confirm this trend when asked about his favourite spaces during a question-and-answer session at the Globe in 2005; most directors, he said, tended to enjoy working in the Cottesloe, the Swan, and the Royal Exchange (he listed the in-the-round spaces at the Orange Tree, Richmond, and at the Globe as his personal favourites), but dislike the Lyttleton, the Birmingham Rep, and the Royal Shakespeare Theatre. In fact – perhaps in response to this sort of bad feeling – the RST is now mid-way through a four-year transformation into a courtyard space, with a thrust stage and a maximum distance of 14 to 16 metres between the stage and the back rows of the audience. In an RSC press release, Artistic Director Michael Boyd explained the reasons for the refit:

Most major new theatres of the last century have moved away from the 'us and them' of the 19th century proscenium 'picture frame' in search of spaces which celebrate the interaction cinema can't achieve. Our commitment to bring an immediacy and clarity to Shakespeare means we need to bring the audience to a more engaged relationship with our actors. The best way we can achieve this is in a bold, thrust, one-room auditorium – a modern take on the courtyard theatres of Shakespeare's day. Actors, directors and audiences alike want a more intimate experience than the current RST can offer.

(RSC 2004)

Interestingly, the temporary theatre being used by the RSC in the interim is also a courtyard theatre, meaning that all three of the RSC's Stratford spaces will from now on be variations on this formation. I wonder whether this will make the RSC's transfers to West End proscenium-arch spaces more problematic than they already are.

If a space is to be a 'collectivist' one, then, it is vital that it seems shared. Proximity to the actors is important: if we are to enter into an imaginative collaboration with them, it is important that we can see them well, and certainly some of the physiological effects of theatregoing – the adrenalin rush, the increased heart rate, the involuntary shudder – are intensified when swordfights and suchlike are breaking out within 'swinging distance' of audience members.[9] But proximity to other audience members is equally important in creating a sense of unity. Unreserved seating is often employed by the Young Vic and by the Théâtre du Soleil, and both theatres, in common with (for example) the Tricycle, the Swan, the Rose Theatre, Kingston, the Orange Tree, Peter Brook's Bouffes du Nord theatre in Paris, and Shakespeare's Globe, provide shared benches for audience members rather than individualised seats. 'Nothing is so unimportant as comfort,' claims Brook; in fact, he argues, comfort 'often devitalizes the experience' (1989: 147). Declan Donnellan appears to agree:

> It is possible for the seats to be a fraction too comfortable. It's not good that people should fall all the way back into their seats. It's quite important, and something that's becoming lost in our culture, that the audience should be invited to lean forward and make the piece of theatre work for them. The audience and the actors should share a communal imagination. There's a huge difference between leaning forward into an event and sitting back and watching it happen.
>
> (Mulryne & Shewring 1995: 105)

Besides, the relationships formed among audience members during the show – what McAuley describes as the 'Spectator/Spectator look' (2000: 264) – constitute an important part of the meaning-making process. When Illyria brought their open-air *Tempest* to Tonbridge Castle in 2004, the presence of focused but noisily laughing children in the audience, who at one point formed a gaggle around the actor playing Trinculo, allowed the older playgoers to share vicariously in their enjoyment. Indeed, as Granville-Barker pointed out in 1922,

> One of the tests of a good performance is the feeling of friendliness it creates among the spectators. When the curtain falls on the first act, and a total stranger turns round to speak to you and you respond without restraint, you may know that the play has achieved one of its secondary – and presumably, therefore, has not failed of its primary – purposes.
>
> (1922: 216)

Nicholas Kent, director of the Tricycle, claims that the theatre's shared benches frequently lead strangers to open conversations with one another – and eventually, in six cases, to marry (Mackintosh 1993: 24).

The sense of 'shared experience' often becomes even more marked when the production takes place in the open air. The effects of shared light mean that, as Illyria's associate producer Jennifer Rigby puts it, 'there are none of the barriers that you have in conventional theatre between the audience and actors' (Teeman 2005). Collective endurance of inclement weather can engender what is often referred to in reviews as a 'Blitz' or 'Dunkirk' spirit among audiences – the collectivism being blurred in these cases, as so often, with nationalistic overtones. Ian Talbot, artistic director of the Open Air Theatre, Regent's Park, describes this democratising effect with reference to an imagined British national character:

> We are a nation of outdoor theatregoers. We love getting rained on, the democracy of it, the feeling that we're all in this together – equality rules.
>
> (Teeman 2005)

As we saw in the last chapter, of course, constructions of communality are always politically loaded.

The kinds of 'collectivist' space described here do, of course, allow a renewed capacity for interplay between *locus* and *platea*, and this in turn might allow for the generation of a more fluid and disjunctive 'critical' attitude. This can be enabled in any number of ways: 'backstage' areas may be visible (as they are in many of Ariane Mnouchkine's productions), or a representation of them displayed (costume rails, for example, or out-of-character actors remaining on stage on side-benches).

Stage exits might be situated through or behind audiences, so that actors transgress a liminal space which blurs *locus* and *platea* (this is particularly the case when audience members have had to traverse the acting space upon entering the auditorium). Gregory Doran's RSC production of *Macbeth* at the Young Vic in 2000 made particularly interesting use of this blurring of *platea* and *locus* when Antony Sher delivered Macbeth's famous metatheatrical speech, identifying himself as the 'poor player / That struts and frets his hour upon the stage, / And then is heard no more' (5.5.23–5). As Brown argues, this moment draws the very real attendance of the audience into the play's 'metaphysical, political, and moral dilemma' (1999: 40), and Sher's performance played upon this tension. Gradually, as the speech progressed, he encroached further and further into the audience's space, before leaving the auditorium through the entrance by which most of us had come in, and delivering the final few words as he went. Due to the shape of the space, the transition from one to the other was far less distinct than it would have been on a picture-frame stage – and, perhaps, more effective for that very reason.

'New Elizabethan' space

An analysis of Shakespeare in 'collectivist' spaces would be incomplete without a discussion of the complex nature of actor/audience collaboration at what might be called the 'new Elizabethan' theatre spaces, and at the new Globe in particular. Architects of many indoor theatres have drawn inspiration from the Elizabethan theatre, perhaps most explicitly at the Swan and more recently at the newly constructed Rose Theatre at Kingston (Figure 6.1). The Rose, based on the shape of the Elizabethan playhouse of the same name, is, according to its Artistic Director Peter Hall,

> intimate yet epic; a place for private scenes or surging battles. ... It also fulfils the mandatory requirement for Shakespeare: it is large enough for the actor to use his voice fully, yet small enough to allow him to whisper. Above all, it is a place for the audience's imagination.
>
> (Hall 2004)

When it was visited by Hall's own production of *As You Like It* for three weeks in late 2004, Michael Billington found that its 'its rough-hewn quality' reminded him of the Young Vic (*Guardian*, 6 December 2004); the *Independent*'s Paul Taylor, meanwhile, noted that the space itself had 'nothing remotely antiquarian or fusty' about it, and that it offered a

Figure 6.1 The view from the circle at The Rose Theatre, Kingston.

'timelessly liberating' relationship between actor and audience. He did observe, however, that the performance itself was 'too playfully low-key and privately self-ironic for the venue' (*Independent*, 8 December 2004). Stratford's Swan Theatre, meanwhile, has played host to many more Shakespearean productions. Michael Reardon designed it in the 1980s as 'a framework within which the actor and audience would exist on equal terms and interact' (Mulryne & Shewring 1995: 25), and Robert Hewison has noted that rather than being a 'slavish recreation' of an Elizabethan or Jacobean theatre, it is in fact a synthesis within one building of the indoor and outdoor playhouses of that period (1995: 54). Like the Rose, it offers a space which is at once 'Elizabethan' in its layout and 'modern' in its feel.

This is not a quality shared by the new Shakespeare's Globe – a theatre which inhabits a unique territory within modern theatre architecture. While it is undoubtedly a 'collectivist' space in many respects, it is also deeply fragmented in many others. The divides within its supposed unity can create a surprising and profound effect. First, of course, there is the striking clash of the new with the pseudo-Elizabethan: the sleek twenty-first-century foyer and piazza lead directly into the reconstruction, which itself juts up into the skyline of London's South Bank beside some overwhelmingly contemporary architecture; inside, electric 'fire exit' signs sit uncomfortably on the carefully thatched roofs while digital cameras flash all around. Even after the performance has started, the presence of backpacks, raincoats, and passing helicopters overhead serve as constant reminders that despite the carefully reconstructed Elizabethan costumes and surroundings, no miraculous time-travel has been undertaken.[10] A second divide is that between the groundlings and the galleries. While the Globe is certainly a participatory space, with often vocal contributions from audience members, it is often striking that there seems to be a void separating the furthest back of the groundlings from the front of the galleries, in which the energy of the performance dissipates. This dead space is perhaps a result of modern health and safety regulations, which do not allow the pit to be crammed with spectators, or perhaps occurs merely when the Globe does not sell all its tickets (indeed, the energy of the Globe space is markedly dampened when it is only half-full). Whatever the cause, a large number of playgoers with gallery tickets will frequently abandon their seats to join the groundlings in the yard for the second half of the play. Kiernan has recorded that this is usually because 'they thought the groundlings were having more fun' (1999: 111).

Another conspicuous discord within the Globe space is its uneasy mix of 'rough' and 'holy'. The space itself, painstakingly reconstructed, galleried like a cathedral, feels almost sacred at times; its visitors, however (and often its performances too) are distinctly 'unholy' (Thomson 2000: 191). Actor Christian Camargo has commented that 'it's like being in a church, but it's a solemnity that gets cracked by the performance' (Kiernan 1999: 153). Certainly Kathryn Hunter's 2005 production of *Pericles* disrupted the 'holy' space of the auditorium as acrobats swung across the space on ropes and ran horizontally, suspended, along the fronts of the galleries. This invasion produced a remarkable effect; naturally the potentially dangerous stunts and the audience's physiological reactions to these located the performers very much in *platea* space, but I suspect there was more to it than this.

Sir Anthony Quayle once defined a successful theatre as being 'half a church and half a brothel' (Mackintosh 1993: 1–2), and Brook, as we have seen, has suggested that the way in which Shakespeare's plays produce their 'disturbing and the unforgettable' effects is though their 'unreconciled opposition of Rough and Holy' (1990: 96). I wonder whether the Globe's uneasy fusion of new and old, of rough and holy, in fact serves to allow movement between what Brook describes as the 'inner and outer worlds' characteristic of Shakespeare's plays. Directing different lines to the groundlings and to the galleries enables the kinds of interplay – by no means unifying – which we identified earlier in this chapter.

Again, it is this 'atonal screech of absolutely unsympathetic keys' (Brook 1990: 96) which produces such an interesting and profound effect.

One final 'new Elizabethan' space which deserves a brief mention, despite having never (to my knowledge, at least) been a venue for Shakespearean performance, is the 'Globe Theatre' at The Bedford, a large pub in Balham. A dimly lit indoor space with the same circular, galleried structure and thrust stage as the more famous Globe, it feels somehow an even more 'collectivist' space (probably due to its location within a pub) than the other 'new Elizabethan' spaces discussed above. Political comedian Mark Thomas filmed his television series *The Mark Thomas Comedy Product* there, and I cannot help but wonder how a 'popular' production of Shakespeare would work in the space. John Russell Brown told me that he had an as-yet-unfulfilled ambition to direct Shakespeare in a comedy club (Brown 2004), and in *New Sites For Shakespeare* he suggests why: comedians, he argues, 'show, in theatrically more limited contexts, how audience-stage reactions can add vitality and, sometimes, a sense of danger and adventure to performance' (1999: 101–2). I feel sure it can only be a matter of time before a Shakespearean company chose to experiment with the space.

'Appropriated' space

So far, we have looked at the ways in which the configuration of a theatrical space might mark it out as a 'popular' one. However, there is far more to a space than its layout, and a subject we have still to address in detail is Lefebvre's concept of 'dominated space'. In a theatrical context, dominated space is certainly not limited to the proscenium-arch theatre. McAuley argues that

> not only the contemporary commercial theatre, dominated by designer and technology, but also the art theatre, controlled by the director and the funding authorities, constitute dominated spaces that to a significant extent have been alienated from their primary users.
>
> (2000: 281)

Michael Elliott, designer and first Artistic Director of the Royal Exchange, expressed a similar sentiment (albeit in less theoretical terms) in a radio broadcast 1973:

> Does the imposing monument encourage the timid playgoer? ... There has been much talk of fun palaces, and perhaps there is a

quarter truth in the idea. The small minority loves going to Covent Garden or the National, for social and nostalgic reasons, as well as theatrical, but perhaps many of us would find more interest and excitement in less conventional surroundings. Tents, halls, gardens, rooms, warehouses.

(1973: 20)

This section, then, looks at the 'appropriation' of such spaces for Shakespearean performance, and at the semiotic and phenomenological effects of this.

The chief category among 'appropriated' sites for Shakespeare is that of converted or 'found' space. Lefebvre argues that every space has an 'etymology', in the sense of 'what happened at a particular spot or place and thereby changed it': as he puts it, 'the past leaves its traces; time has its own script' (1991: 37). When it comes to theatre space, its 'etymology' can (and indeed must) have a profound effect on the experience of the theatrical event. Cliff McLucas of the site-specific performance company Brith Gof suggests that

deciding to create a work in a 'used' building might provide a theatrical foundation or springboard, it might be like 'throwing a six to go', it might get us several rungs up the theatrical ladder before we begin.

(Kaye 1996: 213)

A 'found' space, McLucas suggests, will have had such a rich multiplicity of signs and symbols inscribed upon it over the course of its existence that it might grant a piece of theatre many additional layers of meaning even before it has begun. It is perhaps a type of space more in keeping with the disjunctive and dialogic impulses of what I have termed the 'critical' model of popular theatre.

'Found' sites do not necessarily have to be non-theatrical spaces. Peter Brook has converted old theatres both in Paris (the Bouffes du Nord) and in Brooklyn (the Majestic), using their theatrical associations – the 'ghosts of past productions', as Mackintosh put it – 'both as images in themselves and as a source of energy' (Mackintosh 1993: 84). Reardon, writing in 1995, notices that designers in recent years 'have abandoned the ideal of "invisibility" and are restoring historic theatres, which of course were never intended to be "invisible"' (Mulryne & Shewring 1995: 25). The creative advantages of visible *platea* space have already been indicated, but science suggests something more: experiments with electrodes connected to the human brain in the late 1970s showed that the 'information rate' of a building – its variety of colour, pattern, and conflicting symbols – determined in part the level of psychological arousal in the subject. Participants spending time in busy, 'festive' spaces apparently 'laughed quicker and cried quicker when exposed to stimuli' (Mackintosh 1993: 81–2), while those in buildings with low information rates were 'understimulated' or even depressed (McAuley 2000: 59). The implications for the foundation of a 'collectivist' theatre space seem obvious; discussing the 'black box' theatre spaces built in the 1960s, however, McAuley suggests that 'it has only been with hindsight that the connection between the information rate of the audience space and the passivity of spectators has been postulated' (2000: 59).

If older theatrical spaces provide a higher information rate, then, it seems fair to assume that this will also be true of spaces not theatrical in origin. Venues such as the Almeida (a Victorian Literary and Scientific Institute), the Orange Tree (an upper room in a pub), the Watermill (as its name suggests, an old mill), and all three incarnations of the Stephen Joseph Theatre have shown that the permanent conversion of non-theatrical into theatrical space can reap rewards long after the original renovations. Jude Kelly feels that the great advantage of spaces such as Glasgow's Tramway Theatre (another space originally converted by Peter Brook, for the world tour of his *Mahabharata*) is that they 'already have a human history in the very bones of the building' (Mulryne & Shewring 1995: 74), while the Tramway's former director Neil Wallace, in what seems directly to echo McLucas's comments, describes the benefits of the many 'found' spaces converted by Brook as follows:

> the innate character, history – one might even call it *drama* of the environment – contributed in some magical way to the theatrical experience of which the audience was a part. … Whether through anticipation, or the thrill of unknowing, or a measured sense of ritual (especially in the outdoor spaces), the pre-performance sensation suggests that, before a single actor has appeared, or a note of music been sounded, or a house light dimmed, the theatrical adventure, the story, has already begun.
>
> (Wallace 1995: 62)

As we have seen, Counsell has suggested that the interplay between abstract and concrete (or *locus* and *platea*) is central to Brook's theatre (1996:163–4), and it would appear that not only the formation but also the 'information rate' of his theatre space has no small part to play in this. Mackintosh relates the story of a National Theatre building committee meeting in the 1960s when architect Denys Lasdun complained that Brook would prefer a Brixton bomb site to anything he could design; Brook, apparently, simply replied, 'Yes' (Mackintosh 1993: 86).

Of course, not all conversions are permanent. Touring companies like Northern Broadsides temporarily inhabit such venues as cattle markets, churches, indoor riding stables, and mills, and on their official website they describe their company style as 'innovative, popular and *regional*' (Northern Broadsides 2006; my emphasis). As their name implies, Northern Broadsides are primarily a northern company, speaking in their natural Yorkshire accents and visiting (in their early years at least) predominantly northern audiences. This sense of regional locality is particularly important in relation to their Shakespearean work, since Shakespeare is all-too-often seen as the domain of the capital, of the south, and of the received-pronunciation middle-classes. Discussing their 1992 production of *Richard III*, Peter Holland argues that

> [a]udiences in the north of England, for whom the production had been conceived, were not required to see Shakespeare as an expression of Home Counties middle-class culture which patronised them.

Northern Broadsides, Holland suggests, 'reclaimed Shakespeare in a piece of cultural annexation that reappropriated high culture and its geographical polarity' (1997: 152). 'Appropriation', here, becomes symbolic as well as spatial.

A sense of 'localness' is one of John McGrath's nine defining features of popular theatre (1996: 54–60). The reasons for this may become clear when we consider the concepts discussed earlier of 'ownership' in relation to theatrical space; there can surely be no greater sense of the audience's collective ownership of a space than when that space is already a material part of their local community. Brith Gof's Mike Pearson describes the way in which his company strives to appeal to audiences in Wales in their own spaces, and on their own terms: 'barns, churches, chapels, what-have-you ... the venues in which a Welsh, particularly a Welsh rural audience, would feel more at ease in' (Kaye 1996: 210). The Rude Mechanical Theatre Company's summertime 'Theatre of the Green' tour travelled mainly to 'village greens and recreation grounds in small rural communities throughout South East England' (Rude Mechanical Theatre Co. 2006), and director Pete Talbot explained to me that

> [b]ecause our shows are pitched at communities, people come not because it's theatre, but because it's a village event. So in fact we've

been pretty successful in getting all kinds of people to our shows who don't normally go to the theatre: farm workers, you know, because they're rural communities, all kinds of people who live in villages who identify with the village and go to village events, but would never go to the theatre. So I think we've been pretty successful, our audiences are more diverse than they appear at first sight.

(Talbot 2003)

Such an appeal to 'localness' and its implied reliance on community and word-of-mouth has perhaps been exhibited more famously by Footsbarn, who since 1975 have erected their distinctive tent in open and often rural spaces all over the world.

There is, however, a sense in which touring can ignore 'localness' and temporarily appropriate a space in the name of elite culture, alienating it from its local audience rather than deferring to their claims of ownership. The RSC, for example, toured to a variety of non-theatrical spaces with productions of *Julius Caesar* and *The Two Gentlemen of Verona* in 2005. Their publicity material describes this 'unique mobile auditorium':

Travelling on five 45 foot articulated lorries and carrying over 50 tonnes of equipment – including everything from the auditorium, seating, costumes and set, to the tea-urn and washing machines – it takes 24 hours to erect the RSC's mobile auditorium, transforming a sports hall or community centre into a state of the art theatre. Once up, the auditorium recreates the intimate atmosphere of a theatre such as the Swan Theatre in Stratford-upon-Avon, providing the ideal setting to watch the company of 20 actors perform these two contrasting plays.

(RSC 2005)

This suggests not so much a collaboration as the distribution of a consumable theatrical product. Discussing the RSC's touring shows ten years earlier, actor Simon Russell Beale admitted that he could see the argument for 'needing to adjust to new audiences and new places, instead of operating the notion of an RSC product which you are taking around' (Mulryne & Shewring 1995: 108).

The difference between casual appropriation by a touring product and collective appropriation by actors and audience alike is an indistinct one, and determined by a variety of factors. This might, perhaps, be illustrated by two very different experiences of touring open-air Shakespeare productions I saw in late summer 2004, one being

Chapterhouse Theatre's *As You Like It* at Michelham Priory, East Sussex, the other, Illyria's *Tempest* at Tonbridge Castle, Kent. First of all, the two locations, while both regional heritage sites, had fundamentally different meanings for their audiences: Michelham Priory is a large site in the middle of the Sussex countryside, accessible only by car, whereas Tonbridge Castle is much more compact, and forms the focal point of the surrounding town. As such, Tonbridge Castle already *represented* its audience to a much greater extent. Secondly, at Tonbridge there was a greater sense of local reference in the performance itself. This might, of course, have been spurred on by the more emphatically 'local' venue, but I suspect that Illyria's style accommodated much more potential for local reference than Chapterhouse's anyway – Chapterhouse's performance resolutely ignored the idiosyncrasies of its particular combination of location, weather, and audience, while Illyria played upon it, picking out audience members and ad-libbing with them. Open-air theatre, of course, provides increased opportunities for local reference, since when it occurs, the accidental presence of stray pigeons, seagulls, cats – even dolphins, according to Phil Jackson, manager of the Minack Theatre, Cornwall (Teeman 2005) – almost demands some kind of acknowledgement by performers.

'Disjunctive' space

As the section above indicates, it may be in appropriated space that the 'critical' model of popular Shakespearean performance finds its most effective expression. As we have seen, site-specific performance will inevitably add a 'metatextual' layer of sorts, picking up on the potential in the multiplicity of conflicting signs present in busy, everyday spaces for creating interesting, fractured, layered performance. Kaye defines site-specific performance as 'practices which, in one way or another, articulate exchanges between the work of art and the places in which its meanings are defined' (2000: 1). Central to site-specific performance is the idea that as a 'text' inscribed upon by human activity, no human-built site can ever be politically neutral or meaningless.[11] This is not to say that resistance is impossible: that which is 'dominated' can be 'appropriated'. Lefebvre's theoretical model suggests that as a 'social product', space is always 'in process' and thus its meaning is constantly in flux and never complete. Site-specific performance thus becomes a form of appropriation.

This is a view shared by Mike Pearson and Cliff McLucas of Brith Gof (Turner 2004: 374). In their work, pieces deliberately do not fit entirely comfortably within their spaces; rather, it is the intertextuality between piece and space which produces the interesting tensions of the performance. Pearson suggests that 'the denotative and connotative meanings of performance are amended and/or compromised by the denotative and connotative meanings of site' (Kaye 1996: 214), while McLucas, using his own, now widely used terminology of 'the Host, the Ghost and the Witness', explains the process as follows:

> The Host site is haunted for a brief time by a Ghost that the theatre makers create. Like all ghosts it is transparent and Host can be seen through Ghost. Add into this a third term – the Witness – i.e. the audience, and we have a kind of Trinity that constitutes The Work. It is the mobilisation of this Trinity that is important – not simply the creation of the Ghost. All three are active components in the bid to make site-specific work.
>
> (Kaye 2000: 128)

It is in the gaps between the three, McLucas suggests, that the meanings of the performance are created:

> There's always a mis-match between the 'host' and the 'ghost', and from the beginning of the work it's fractured, it's deeply, deeply fractured … they are more discursive, and have gaps in them – you can see other things through.
>
> (McLucas et al. 1995: 51, cited Kaye 2000: 56)

Applied to Shakespearean performance, McLucas's model suggests a 'disjunctive' use of space which might lend itself to metatextual commentary.

This might make us think again about those productions of Shakespeare which claim to be 'site-specific'. Creation Theatre tops this list (Creation Theatre, 2006), with its prolific output of what its official website describes as 'site-specific theatre in unusual locations' (these include the BMW Group Plant, Oxford, an elaborate Spiegeltent, and Saint Augustine's Abbey, Canterbury). While these locations undoubtedly add a great deal to what might be described as the 'atmosphere' of the performance (more specifically, the set of semiotic associations prompted by the location which impact upon the audience's meaning-making process during the performance), I am not sure that Creation's productions interact with their locations in quite the same way as Brith Gof's. Even productions which deliberately situate themselves in locations which are as semiotically 'busy' as is possible, such as Frantic Redhead Productions' now-annual walkabout *Macbeth* at the Edinburgh Fringe Festival (officially sold-out for seven years running), can often seem to be avoiding confrontation with their environment. Frantic Redhead's production takes audiences on a promenade tour of various locations in Edinburgh's city centre, in each of which discrete scenes of *Macbeth* are played out. Baz Kershaw suggests that street theatre of this sort will inevitably be 'thoroughly contaminated by its wider cultural context' (Kershaw 1999: 7), but Frantic Redhead seemed perversely determined to overlook this contamination. Occasional references to the 'iron horses' on Edinburgh's roads and an instruction to 'wait for the appearance of the Green Knight' at the pedestrian crossing were made as the audience walked between locations, but during the scenes themselves the performers acted as if hermetically sealed from the contemporary city. This became uncomfortably apparent in one scene as the entire company tried desperately to ignore the presence of a local boy making cheeky comments in the centre of their acting area. The touring Out Of Joint production discussed earlier in this chapter might be considered a more successful 'site-specific' *Macbeth*, but in sealing itself from the outside world within private buildings, it did not have to confront the same mass of distracting signs faced by Frantic Redhead. The reluctance of these productions to embrace such distractions can, I think, be attributed to a desire to present the texts 'faithfully' and with respect for their unity. It is a priority which is fundamentally at odds with the 'critical' attitude and indeed with 'site-specificity' in general (or at least the definition of site-specificity put forward here).

It strikes me that there is a largely unfulfilled potential for a popular deconstruction of Shakespeare in the often-chaotic setting of open-air theatre (though early and untheorised attempts have arguably been made by companies such as Ophaboom, Illyria, and Oddsocks). Given the tentative definition of 'disjunctive' theatre space suggested above (that is, embracing as wide a variety of conflicting symbols as possible), one might be able to make the case for Shakespeare's Globe as the most potentially 'disjunctive' of London's dedicated theatrical venues: as we saw earlier, mock-Elizabethan features sit side-by-side at the Globe with intrusions from contemporary reality. While other spaces strive for invisibility, the Globe advertises its status as a theatre and as a reconstruction, entailing a necessary element of parodic self-reference by performers there. Mulryne and Shewring, writing before the Globe first opened its doors to the public, identified in its fellow 'new Elizabethan' theatre the Swan 'a conscious awareness of the theatricality of theatrical space' (1995: 12).

The Globe, however, goes even further. In a sense, not only is the original building 'quoted' in the reconstruction, but as we saw in Chapter 5, mediatised quotations of the original building are, in a sense, 'quoted' too. Brown comments that

> [w]hen we go to the new Globe, we do not travel backwards in time but enter a little, carefully fabricated world with the rarity, pretence, and educational advantages of an Elizabethan theme park.
>
> (1999: 190–1)

His analysis is probably intended as disparagement; it recalls Baudrillard's notion of 'hyperreal simulacra', and suggests that like Disneyworld, the Globe might be argued to be a simulation of a place which never really existed (in this form, at least) in the first place (see Baudrillard 1988). But as a space which is ambiguous, heterogeneous, fragmented, and filled with simulations of the past – qualities celebrated by one of the godfathers of postmodern architecture, Robert Venturi, in his ground-breaking *Learning from Las Vegas* (1978: 118) – it offers the potential for dynamic shifts between critical and collectivist modes of spectatorship, between 'textual' and 'metatextual' attitudes. As Holderness argues:

> Rising in extraordinary architectural isolation among the dereliction and tower-blocks, a reproduced Elizabethan theatre could represent a triumph of post-modern style, capable by its pastiche and quotation of calling attention simultaneously to present and past, a contradictory synthesis of modernity and the antique. The relationship between building and location would be one of shocking incongruity rather than any smooth absorption of a distinctive local patina. Stripped of all pretensions to authenticity, could not such a building offer the spectator a critical consciousness of cultural appropriation?
>
> (2001: 99–100)

It is a potential which Holderness argues shows no signs yet of being realised. But as he suggests, it is there.

Personal Narrative 7
'It's the famous bit!': Fragments of *Romeo and Juliet*

Today we can experience *Romeo and Juliet* only in fragments.

Perhaps it's always been true, and perhaps it's true of all fiction, but it's especially true today, and it's especially true of *Romeo and Juliet*. A *Romeo and Juliet* today is a post-Zeffirelli *Romeo and Juliet*, a post-Baz Luhrmann one, post-Dire Straits, post-*West Side Story*, post-*Shakespeare In Love*. It's one which has been splintered into pieces by images of finger-clicking gang members, of cartoon animals re-enacting the balcony scene, of quotations and misquotations in commercials, sketches, textbooks, and sitcoms. I directed a production of the play for the Pantaloons theatre company with this in mind, and afterwards, in attempting to document my experience, I realised that only a similarly fragmented account could even begin to describe the multifaceted jumble of a *Romeo and Juliet* that emerged from the twenty-odd performances we did, in five different locations, at various times of day and night, in wildly differing circumstances. This is an account of those performances, compiled from reviews, recordings, audience feedback, cast anecdotes, and, of course, my own memories.

Steering clear of Leo

We are rehearsing 5.1, where Romeo learns of Juliet's apparent death. Dom Conway, playing Romeo, is finding it difficult to respond to the 7-line speech by Balthasar (Tom Hughes) in which the news is revealed to him. We try playing it so that Romeo embraces Balthasar to greet him, but the embrace becomes a cling for physical support as he learns of Juliet's fate. We reach Romeo's next line: 'Is it even so? Then I defy you, stars!' (5.1.24). Dom breaks away. He wants to try something more understated. He's anxious to avoid playing the line like Leonardo DiCaprio.

Wherefore art thou Romeo?

We rehearse the 'balcony scene'. Clare Beresford plays Juliet as a young girl fantasising about true love; before she sees Romeo, she's play-acting at being in love, sighing and quoting 'Ay me!' (2.1.67) simply because they are the kinds of things lovers are *supposed* to do (Mercutio's ironic quotation of the same phrase at 2.1.10 implies as much). Romeo will sit among the audience to watch her. Dom plays Romeo's first two speeches with great energy, building in intensity and anticipation: the natural climax seems to be Juliet's 'O Romeo, Romeo, wherefore art thou Romeo?' (2.1.75).

But Clare is having trouble with this line. 'It's just so famous!'

Remembering Peter Brook, I decide that a climax of laughter may be appropriate here.

When the scene is finally performed, Romeo sits on a grassy bank among audience members. 'It is my lady,' he says smugly, pointing Juliet out to one of the nearby spectators. 'O, it is my love,' he says, grinning at another (2.1.53). He is confidential, like a schoolboy with a crush pointing out the object of his affections to a group of friends. 'See how she leans her cheek upon her hand,' he says, indicating proudly (2.1.65).

'Ay me!'
'She speaks!'

The audience laugh. Romeo's excitement is almost orgasmic as he throws himself to the floor in ecstasy:

> O, speak again, bright angel; for thou art
> As glorious to this night, being o'er my head,
> As is a winged messenger of heaven
> Unto the white upturned wond'ring eyes
> Of mortals that fall back to gaze on him
> When he bestrides the lazy-passing clouds
> And sails upon the bosom of the air.
>
> (2.1.68–74)

'O Romeo!' she suddenly cries.

He sits bolt upright, and without missing a beat, yells 'It's the famous bit!'

A performance transcript

Romeo and Juliet are about to be married.

> FRIAR LAURENCE.   Come, come with me, and we will make short
>   work;
>   For, by your leaves, you shall not stay alone
>   Till holy church incorporate two in one.
>   *He turns to the audience.*
>   Now I suppose you all want to see a wedding, don't you?
>   *The audience responds, mostly 'Yeah!'*
>   Well, Shakespeare didn't write one.
>   *Laughter.*
>   But you've all seen the film. Well, we don't have Leonardo
>   DiCaprio, we don't have Claire Danes, we don't have a cathe-
>   dral, and we certainly don't have a small singing boy.
>   *Laughter.*
>   But what we do have is a large amount of confetti, and a will-
>   ingness to ad lib.

Reviews

We're getting some very encouraging feedback from the public and the press, but some of it seems to be missing the point. 'There's just enough panto to keep kids occupied, but enough proper theatre for the Shakespeare fans,' says an audience member on the edfringe. com forum. A five-star review from *Three Weeks* delights us all, and we jokingly suggest quoting an endorsement on our posters in order to attract the 'MTV generation': 'even small children or those with a short attention span will be captivated' (*Three Weeks*, 13 August 2006).

Some thoughts from the cast

Dom Conway (Romeo) on improvisation:

Because the audience changes and the space changes so hugely from show to show, there is very little sense of having to rely on previous discoveries. There is no archetypal performance of the play that we try to recover or stray from; each performance is a self contained exploration into the play and the type of theatre we are engaged in. ... My favourite moments were the stage invasions and the first-time improvisations. Those moments were so exciting because they were intense moments of anarchic discovery often prompting even more exciting repercussions throughout the play and further shows.

Dave Hughes (Friar Laurence) on the audience's willingness to laugh:

Because they were so ready to laugh, but at the same time, ready to get the tragedy, it was almost like we were pushing it as far as we could. And that was why I likened it to being on a thread that could snap, because if you pushed it too far, then it would all fall to pieces. But the audience were willing to be pushed just that little bit extra, and it felt slightly chaotic, but at the same time, just really good fun.

Clare Beresford (Juliet) on the comedy:

I did worry that it might take away some of the sadness from the ending; but I was completely wrong in my initial judgement, as it in fact did the complete opposite. I didn't completely realise how well juxtaposing the two, without fear, to the very end, would work. ... [An audience member] described how he'd made a personal choice to focus on the sadness, which continued right through the comic points. In fact, he even pointed out that the audience members who appreciated the comedy towards the end upset him more, as he couldn't understand how they couldn't see what he was seeing, two lovers dying to be with one another. And therein, I believe, lies the evidence that what we were trying to achieve does work, as the audience members who choose to read into the comedy until the very end have the freedom to do so, not even realising that they are adding to the anguish felt by the other members who are already locked in by the tragedy.

Panto

Perhaps the reviewers have a point. We open with a panto-style sing-off, in which the audience is divided into two by the Capulet and Montague servants. 'Wait, wait!' cries a cast member. 'We're supposed to be doing Shakespeare: this is turning into a pantomime!'

'Oh, no, it isn't!' reply the cast in unison.

Later, a desperate Juliet, bullied and blackmailed by her family into marrying Paris, will exit the stage to visit the Friar, 'to know his remedy'. 'If all else fail,' she concludes as she leaves, 'myself have power to die' (3.5.242).

Dave Hughes enters opposite as Friar Laurence. '"If all else fail, myself have power to die"', he repeats. 'Our Juliet, there, always the practical thinker.'

He asks the audience if they have any better suggestions. Over the course of our 21 performances, their suggestions will include a cunning ruse involving fake poison ('I might give that a go,' says the Friar); hiding ('Hold, daughter! I do spy a kind of hope,' he later says to Juliet (4.1.68); 'How good are you at hiding?'); various murders (Juliet's father, Juliet's mother, Paris, or in one memorable instance, 'everybody'); and 'go on *Trisha*'.

A small child suggests, 'Go to space.'
'How will she get to space?' says the Friar.
'On a space hopper,' replies the child.
The Friar nods. 'It seems so obvious.'

7
Shakespearean 'Samples'

> I confess myself utterly unable to appreciate that
> celebrated soliloquy in *Hamlet*, beginning 'To be or not
> to be', or to tell whether it be good, bad, or indifferent,
> it has been so handled and pawed about by declama-
> tory boys and men, and torn so inhumanly from its
> living place and principle of continuity in the play, till
> it is become to me a perfect dead member.
>
> (Lamb 1963: 22–3)

Shakespearean simulacra

Nearly 200 years ago, Charles Lamb recorded a phenomenon of
Shakespearean spectatorship which has only gained in currency over
the time since. Already, even then, the constant cultural recycling of
Shakespeare's speech had alienated it from any literal meaning the text
might once have had.

A recent adaptation of *A Midsummer Night's Dream* screened as part
of the BBC's *ShakespeaRe-Told* season (2005) might illustrate the trend
in more contemporary terms. Along with its companion adaptations
(*Macbeth*, *Much Ado About Nothing*, and *The Taming of the Shrew*),
this programme lifted the main characters and central plotlines of
Shakespeare's original and relocated them to a modern setting. Like
some of the others (most notably *Much Ado About Nothing*), it also made
use of both direct and paraphrased quotation from Shakespeare. In the
scene in which Oberon commanded Puck to fetch him the 'love-juice',
for example, Puck employed the idiom of a contemporary Mancunian

drug-dealer ('How much are we talkin'?'), while Oberon's speech made use of Shakespeare's own blank verse (2.1.164–8):

OBERON.   Remember that stuff I used the night I heard the
    mermaids sing?

PUCK.   Those mermaids! Aw, man, we 'ad a wobble on that night!

OBERON.   All in maiden meditation, fancy-free, yet marked where
    the bolt of Cupid fell.

PUCK.   I can score ya for love-juice. No danger.

OBERON.   It fell upon a little western flower –
    Before, milk-white; now, purple with love's wound –
    And maidens call it love-in-idleness.

PUCK.   D'ya want the bleedin' love-juice, or what?

The missing 'I' from 'yet marked where the bolt of Cupid fell' renders Oberon's sentence rather indecipherable, but clearly the literal meaning of Shakespeare's passage is not what matters here: the script is playing with the contrast between what in Chapter 3 I called 'official' and 'unofficial' registers, and Shakespeare's verse signifies by sheer dint of being 'Shakespeare'. Later, Oberon's reconciliation with Titania made use not only of lines from the original text, but also of a compilation of excerpts from Sonnets 39 and 56 and *Romeo and Juliet* (2.1.175–7):

TITANIA.   I dreamt I was in love with an ass. Then I woke up. And
    what do you know, I still am.

OBERON.   I'm sorry. I love you. And you were right.

TITANIA.   What's happened? Why are you being like this?

OBERON.   What if I were to tell you: no more stupid rage?

TITANIA.   I'd say it was a start.

OBERON.   No more commands.

TITANIA.   I never listen to your commands anyway.

OBERON. No more jealousy.

TITANIA.   Why should I believe you?

OBERON.   Because I've seen the damage it does.
O, how thy worth with manners may I sing
When thou art all the better part of me?
Sweet love, renew thy force. Be it not said
Thy edge should blunter be than appetite;
My bounty is as boundless as the sea,
My love as deep. The more I give to thee
The more I have, for both are infinite.

TITANIA.   How I love thee! How I dote on thee!

This medley of Shakespearean snippets once again produces something of a non-sequitur, in this case after 'Thy edge should blunter be than appetite', but again, the audience is not primed to look for textual meaning: Oberon is simply 'doing Shakespeare', and this characterises him as, in this instance, sensitive, romantic, and poetic. It is quite literally a pastiche: a Shakespearean simulacrum. The adaptation illustrates the crucial cultural difference between performing Shakespearean text and performing 'Shakespeare' the cultural icon.

Contemporary Shakespearean performance is, whether it wills it or not, a similar sort of pastiche. Simply by merit of its being 'Shakespeare', it taps into a dizzying array of myths and associations, many of them (as we have seen throughout this book) mutually contradictory. 'Shakespeare' has accrued a set of cultural meanings which are quite separate (but not entirely *separable*) from the literal meanings of the Shakespearean text: Graham Holderness, and before him Terry Eagleton, distinguish between, on the one hand, Shakespeare the man and his playtexts, and on the other, 'Shakespeare' the myth (Holderness 2001: x; Eagleton 1988: 204). It may be that as the works recede from view, the myth steadily grows in mass. Gary Taylor has shown that while acquaintance with the plays is disappearing from popular consciousness, 'Shakespeare' the cultural icon continues to accrue cultural capital (Taylor 1999: 197–205); in his earlier study *Reinventing Shakespeare*, he concluded by comparing Shakespeare with a black hole:

Shakespeare himself no longer transmits visible light; his stellar energies have been trapped within the gravity well of his own reputation. We find in Shakespeare only what we bring to him or what others have left behind; he gives us back our own values. And it is no use pretending that some uniquely clever, honest, and disciplined critic can find a technique, an angle, that will enable us to lead a mass escape from this trap. If Shakespeare is a literary black hole, then nothing that I, or anyone else, can say will make any difference. His accreting disk will go on spinning, sucking, growing.

(1990: 411)

In performance, as Taylor's analogy implies, disengagement from the various cultural meanings which have been heaped upon Shakespeare grows ever more impossible. We cannot, however hard we try, wrench ourselves free from the myth.

Throughout this study, I have maintained a dual focus on approaches towards Shakespearean performance which I have characterised respectively as 'textual' and as 'metatextual'. These correspond, perhaps, with W. B. Worthen's formulation of 'literary' and 'performative' attitudes towards the theatrical text. For Worthen, a 'literary' (textual) perspective 'takes the authority of a performance to be a function of how fully the stage expresses meanings, gestures, and themes located ineffably in the written work' (1997: 4). Such an approach may not, as we have seen, be 'textual' or 'literary' in the strictest senses of the words: it will often transform, translate, rearrange, or otherwise depart from the Shakespearean script. But when it does so, it does so in the understanding that it is fulfilling – 'surrogating', to use Roach's term – a theatrical function which is inherent in the original text. Under a 'literary' understanding, to use Worthen's example, a 'kathakali *Othello* is said to "work" because classical Indian dance drama somehow replicates the intrinsic dynamics of Shakespeare's play' (2003: 117). Popular Shakespeare, even at its most iconoclastic, frequently claims to be taking us back to the 'real' Shakespeare (see Anderegg 2003 on Luhrmann's *Romeo + Juliet*, for example). Performance innovations are validated with reference to 'original' theatre practices; even radical critics find themselves referring to the kinds of staging which are 'natural to the Elizabethan theatre' (Holderness 2001: 46).

The textual approach, therefore, remains bounded by that which is deemed 'Shakespeare'. Brook advocates a 'healthy double attitude' towards the Shakespearean text, 'with respect on the one hand and disrespect on the other'; he feels that in some Shakespeare plays, 'you can move words and scenes,' but, he notes, 'you have to do it in full recognition of how dangerous it is' (1989: 95). Shakespeare may be altered, adapted, and rearranged: but it is 'dangerous'. The danger is illustrated, perhaps, by those productions considered throughout this book which were censured in newspaper reviews as 'not Shakespeare' (and the role of theatre criticism, as Sinfield puts it, is often 'to police the boundaries of the permissible'; 2000: 187), or condemned by press and academic critics alike – as in Billington's review of Kneehigh's *Cymbeline* – as having revealed 'nothing new about the play' (*Guardian*, 23 September 2006). Rob Conkie's survey of the critical reception of productions at Shakespeare's Globe – a theatre which, after all, is founded upon the assumption that it will deliver new insights into the 'authentic' staging of the plays – reveals that 'Original Practices' productions have on the whole been reviewed far more warmly than more 'performative' productions, such as Brazilian street theatre troupe Grupo Galpão's carnivalised *Romeu e Julieta* (which once again, for Billington, taught 'very little new about Shakespeare's play'; *Guardian*, 15 July 2000). Such critical responses, Conkie argues,

> work, probably unconsciously, to maintain the hegemonic structure of English, authentic, representational, and internally consistent Shakespeare as opposed to and hierarchically above Other, hybridised, presentational and performatively constructed Shakespeare.
>
> (2006: 35)

It is another example, if another were really needed, of what Sinfield has called 'the importance of being Shakespeare' (2000: 185).

Such priorities are generally rejected by a metatextual or 'performative' approach towards Shakespeare in theatre, which typically challenges the authority both of the text and of the myth which surrounds that text. As opposed to the literary perspective, argues Worthen,

> a 'performative' sense of dramatic theatre enables a sharper, dialectical reading of the ways in which different theatrical practices transform Shakespearean writing into action and acting, into meaningful behaviour onstage.
>
> (2003: 117)

The metatext, as we have seen, signals its disjunction from both Shakespeare and 'Shakespeare', and in doing so, offers either implicitly or explicitly a critical commentary upon it. It indicates that performance ('particularly,' notes Worthen, 'the performance of plays in the classical repertory') should be evaluated in its own terms, 'without recourse to an understanding of the perdurable text' (2003: 12). In this sense it might be understood as analogous to Barthes' ideal of a 'writerly' text. As opposed to a 'readerly' text, which is 'like a cupboard where meanings are shelved, stacked, safeguarded', the 'writerly' text will encourage its reader to become an active participant in the creation of its meaning – 'no longer a consumer, but a producer of the text' (1974: 200–1 and 4). The writerly text – or performance, in this case – does not construct itself as the source of a stable meaning; its audience must engage with it, actively, in order to generate a multiplicity of meanings.

Brecht compared the strategy with the use of footnotes, or 'the habit of turning back in order to check a point', calling it an 'exercise in complex seeing' (1977: 44). Bakhtin similarly distinguished between 'dialogic' and 'monologic' texts; Weimann identifies a distinction between the modern, text-centred concept of 'acting', and the pre-Renaissance tradition of 'playing', in which, like *A Midsummer Night's Dream*'s Starveling, actors 'disfigure' the text as they 'present' it (2000: 131–2). Citing both Weimann and Brecht, Terence Hawkes describes what I have called 'metatext' in similar terms:

> [T]hose stratagems whereby the player, ever trawling the audience for a fruitful response, may produce a special kind of 'deformative' effect by somehow adding to, interfering with, or 'bending' the text so that it begins to impart a different, perhaps almost contradictory sense to that which it overtly proposes.
>
> (2002: 112)

Hawkes suggests that 'an appropriate modern analogy' might be found in jazz music (2002: 112). Indeed, he concluded his earlier study *That Shakespeherian Rag* with the same metaphor, suggesting that it provided a sense of text as a site for 'conflicting and often contradictory potential interpretations, no one or group of which can claim "intrinsic" primacy or "inherent" authority' (1986: 117):

> For the jazz musician, the 'text' of a melody is a means, not an end. Interpretation in that context is not parasitic but symbiotic in its relationship with its object. Its role is not limited to the service, or the revelation, or the celebration of the author's/composer's art. Quite the reverse: interpretation constitutes the art of the jazz musician.
>
> (1986: 117–8)

It is an analogy which has been picked up in Shakespearean scholarship more recently by Andrew James Hartley: 'authentic' Shakespeare is impossible, he argues, because 'theatre, like jazz, authorizes itself' (2005: 61).

To sum up, then: when popular Shakespeare productions assume a 'textual' or 'literary' attitude towards the text, they tend, broadly speaking, to imply that popular Shakespeare equals 'authentic' Shakespeare; when they assume a 'metatextual' or 'performative' attitude, the implication may be that popular Shakespeare is a subversion of – or at least a jazz-like 'variation' upon – the 'authority' of the text (Shakespeare certainly provides unique opportunity for a very conspicuous transgression from textual authority – what author, after all, is more 'authoritative' than Shakespeare?). Both of these perspectives, however, are simplifications: I am not sure either fully describes the complex amalgamation of citation and simulation which can be identified in the *ShakespeaRe-Told* adaptation, and indeed in many of the contemporary pop appropriations studied throughout this book. The question of a fixed relationship between 'Shakespeare' and popular performance is ultimately nonsensical: as this study has shown, both 'Shakespeare' and the 'popular' are nebulous cultural constructions, ever-shifting. As Richard Burt suggests in an essay on what he calls 'post-popular Shakespeare',

> there is no authentic Shakespeare, no 'masterpiece' against which the adaptation might be evaluated and interpreted ... there is no longer a Shakespeare icon, a token of Western imperial power, out there to subvert from the margins; the center is already decentered, the original is already hybrid, the authentic is already a simulacrum.
>
> (2003: 17–18)

It is here that the binary model of 'text' and 'metatext' begins to unravel. The 'Shakespearean text', as we saw at the beginning of this chapter, is not simply a written artefact: indelibly attached to it is a bundle of ever accumulating, politically loaded, mutually contradictory myths. There is, in contemporary performance, no stable 'text' upon which variations may be performed. Variations may be performed only on *other* variations – hegemonic ones perhaps, and more widely circulated, but 'variations' nonetheless. In a very significant sense, any performance, whether 'literary' or 'performative', might be understood to be *composed* of such variations: composed, in other words, of metatexts.

The idea of a 'metatext' serves, paradoxically, to recentralise the 'text'; as the term 'metatext' itself implies, a performative approach is still primarily defined by its relationship with, even if it is an antagonism towards, the 'text' (we saw in Chapter 4, for example, that parody often reaffirms the authority of 'straight' Shakespeare through the very act of transgressing it). In Worthen's argument, both literary and performative approaches finally 'share an essentialising rhetoric that appears to ground the relationship between text and performance' (1997: 4). It may be useful here to return to Shomit Mitter, whose study of Brecht was the source of my terminology:

> [W]hereas metatextual pointers depend entirely upon the text for their existence, conventional text is autonomously articulate. Whereas metatext, as marginalia, can be marginalised, text, as that which is, compels attention.
>
> (1992: 65)

Mitter suggests that instead of a primary/secondary 'text'/'metatext' hierarchy, one might present 'two fully realised texts within the body of a single work which comment upon each other through the force of their revealed incongruity' (1992: 65).

Perhaps contemporary pop music provides an analogy more suited to popular Shakespeare in the twenty-first century. In 'Shakespeare for the MTV generation', perhaps – in the words of the 'hip hop poet' Oscar Kightley – 'Shakespeare is like James Brown. ... Shakespeare is someone to be appropriated and sampled' (Cartelli 2003: 186). In sampling, fragments of earlier recordings, musical or otherwise, are incorporated into new recordings, either by direct quotation or by transformation (or both): Mark Katz defines it as 'the digital incorporation of any prerecorded sound into a new recorded work' (2004: 138). Tempo and pitch can be altered; sounds can be repeated, fragmented, layered, played backwards; the sample can be recontextualised or juxtaposed with other sampled sounds. As Kightley's analogy suggests, sampling is (in theory at least) blind to cultural hierarchies. When Shakespeare, then, is the subject of 'sampling' in performance – as opposed to a jazz-like 'variation' – the conglomeration of cultural meanings we know as 'Shakespeare' becomes decentred. The result is pastiche, a bricolage of cultural artefacts, of ironic quotations from 'metatexts' as well as 'texts' (indeed, jazz itself is a frequent source of sampling in contemporary music) alongside samples from pop culture and from the contemporary world at large.

Such a model points the way towards a hybridised popular Shakespeare in which – in what Robert Shaughnessy calls 'a postmodern blanking of both traditional Shakespearean cultural authority and the politicised theatrical agendas that have attempted to contest it' (2002: 187) – no cultural source is privileged above another. Shaughnessy's description is of the experimental theatre company Forced Entertainment's *Five Day*

Lear (1999), a performance which he invokes at the conclusion of *The Shakespeare Effect* as an example of

> the potential for performed Shakespeare to encounter vocabularies – and politics – of performance which call for a far more drastic dismantling of long-standing protocols of textual and theatrical authority than those attempted in the epoch of modernism.
>
> (2002: 196)

The show was a fragmented collage of extracts from *King Lear* (which the company performed with characteristic disjunction, donning at one point second-hand 'Shakespearean' theatrical costumes), with other 'variations' upon it (Donald Wolfit's recording was played, as was company member Mark Etchells' half-remembered retelling of its plot), and other cultural scraps (cocktail lounge music, the song 'Good Night' from the Beatles' *White Album*, and – most importantly for an explicitly political reading of the piece – news of the fallout from the recent NATO air strikes in the former Yugoslavia).[1]

Currently, argues Shaughnessy at the beginning of his book, 'the Shakespearean theatre's relation to technology is shaped by its continuing commitment to verbal integrity and unproblematised actorly presence' (2002: 10). As I hope this study has shown, however, this mythologising of 'authentic' Shakespearean authority is being increasingly challenged by popular appropriations, which sample Shakespearean retellings in their mediatised as well as their theatrical forms (after all, as we have seen, they are culturally inseparable). The cinematic 'authentic-popular-Shakespeare' paradigm described in Chapter 5 may in fact be on the wane already: one of the most recent film representations of a theatrical Shakespearean audience is in the 2007 comedy *Hot Fuzz* (dir. Edgar Wright), which depicts an awful amateur-dramatic live re-enactment of Luhrmann's *Romeo + Juliet*. The sequence playfully deconstructs the ways in which live theatre, in this age of movie-to-stage-play adaptations, is frequently modelled on its cinematic simulacra. No *communitas* here – the audience are evidently horrified.[2]

A theatrical model might be provided by the Wooster Group. A hallmark of this company's work is the juxtaposition of 'classic' texts with pop culture phenomena: *LSD (Just the High Points)* (1984) juxtaposed Arthur Miller's *The Crucible* with the writings of Timothy Leary; *House/Lights* (1999) set Gertrude Stein's *Doctor Faustus Lights the Lights* against the 1964 film *Olga's House of Shame*; *To You, The*

Birdie! (2002) enacted Racine's *Phèdre* concurrently with a game of badminton. The group's first foray into Shakespeare opened in New York in 2007 – a production of *Hamlet* which offered, in the words of Matthew J. Bolton,

> not a new interpretation of the play, but a painstaking (and, at times, cringe-making) excavation and reconstruction of a previous one, a filmed stage production starring Richard Burton.
>
> (2007: 84)

This was Shakespearean sampling at its most literal. On the stage, one large screen and three smaller monitors displayed edited and digitally manipulated clips from the 1964 film (itself an adaptation of John Gielgud's Broadway production) alongside live, streamed images of the performances taking place on the stage. Sometimes the actors onstage mimicked the film, shifting their wheeled chairs and scenery to represent the movements of the camera, or jerking as the film stock degraded. At other moments, they would fill in the actions which were taking place off-screen. Frequently, they would play with the possibilities offered by recorded performance: scenes both live and recorded were 'rewound' and replayed, and electronic effects were employed both to blend the live actors' voices with those of the film, and to distort them. At one point, where a segment of the 1964 film had been lost, a section of Kenneth Branagh's 1996 film was spliced into the edit in its place.

Like Forced Entertainment, the Wooster Group stage interactions between a jumble of cultural fragments. Bolton found *Hamlet*'s shiny leather costumes and plethora of television screens reminiscent of 'a dated vision of the future', echoing such 80s sci-fi as *Max Headroom*, *Dune*, and *Videodrome*; 'perhaps,' he reflected,

> this recycling of pop-cultural visual tropes was intended to make some larger point about the collision of high culture with low, but I found the overall effect to be muddled rather than meaningful.
>
> (2007: 85)

But the muddle is the point. Director Elizabeth LeCompte characterised the group's recent production of *La Didone* (which played Francesco Cavalli's 1641 opera onstage alongside a simultaneous re-enactment of the 1965 sci-fi B-move *Planet of the Vampires*) as a rejection of 'the old cultural distinctions between high and low art':

I know they are there, because people say so, but when I am working I do not feel that I am making fun of one or the other. I think it is all the same. What I am aiming for is styles clashing up against each other. I am not looking for the high or the low; I am looking for the dialectic and finding the synthesis.

(Fisher 2007)

Claims of a complete rejection of cultural hierarchies must be seriously questioned: the production was sold and advertised as *La Didone* rather than as *Planet of the Vampires* or under another title, and its irony was much more evident in its quotations of the B-movie than in its treatment of the opera. But LeCompte's choice of words is telling – her aim is for both 'the dialectic' *and* 'the synthesis'. It illustrates the simultaneous and self-contradictory impulses towards both disjunction and assimilation embodied by popular Shakespeares throughout this book.

In this sense, where Shakespeare and popular culture appear most opposed, they might better expose one another: veneration might be undercut by irreverence, mythologising by parody, the impulse towards coherence through fracture. But where they simultaneously overlap, the fixed nature of this binary might be thrown open to question.

Personal Narrative 8
Rough Magic

Tonbridge Castle, August 2004. We were nearing the end of Illyria's open-air production of *The Tempest*, and I was sitting on a damp rug beneath a clear, starry sky. I'd drained the last of the tea from my thermos flask, my hands were a little cold, and my breath was catching on the night air. On the stage before me, Prospero, propping himself up with crutches (for most of the production he had been wheelchair-bound), was articulating his promise to surrender his magical powers:

> But this rough magic
> I here abjure. And when I have required
> Some heavenly music – which even now I do –
> To work mine end upon their senses that
> This airy charm is for, I'll break my staff,
> Bury it certain fathoms in the earth,
> And deeper than did ever plummet sound
> I'll drown my book.

$$(5.1.50–7)$$

The metaphor was not a new one, not to me, nor to Shakespearean audiences in general, but it struck me then with some force: there had indeed been a kind of 'rough magic' to Illyria's performance that night. The five actors who formed the company did everything: not only playing all the various roles of the play, but accompanying themselves with live music, running their own stage management, even ushering in the audience and selling programmes at the beginning of the show. They had taken a while to win me over – the opening act had dragged a little, I'd felt, restricted perhaps by its relatively stationary Prospero – and the bright evening sunlight had rendered much of the play's early sequences of magic disappointingly flat and washed-out. But somehow, as the clutter of discarded props at the side of the stage piled up, a different kind of magic took hold.

Over the course of the performance, the sunlight had fallen away, leaving a sharply demarcated floodlit circle around the stage; beyond that, the grounds of Tonbridge Castle were immersed in darkness. One tends not to notice the encroachment of night, and in much the same way, I had not noticed my own transition from disenchantment to enchantment. Almost imperceptibly, the production had cast its spell upon me. The brilliantly executed physical comedy of the scene in which Trinculo and Caliban became the 'monster of the isle with four legs' (2.2.65), moving under their gabardine as one huge body with limbs in all the wrong places – then plucking back the cloak to reveal Ariel tucking into a packet of crisps previously stolen from an audience member – had driven me, and the rest of the audience, to peals of laughter. Trinculo's often-improvised direct address had earned him a following of small children, who had crowded lovingly around him when he came to sit among the audience, and I (and, I suspect, many of the adults in the audience) had shared vicariously in their pleasure. In a wonderful accident, just as Prospero's 'insubstantial pageant' had 'melted into air, into thin air', it had been accompanied by the sight of the coloured flashing lights of a plane flying silently across the sky behind it (4.1.150–5).

Then came the point at which the production's 'rough magic' reached its zenith. After a long build-up, the moment had come for Prospero to fulfil his promise of Ariel's release, and for the play to end. Only Prospero and Ariel remained onstage – the other three actors of the company stood behind them, singing unaccompanied, in close harmony. Prospero was on his crutches; Ariel was a tall, bare-chested actor, costumed in ragged leggings of grey, silver, white, and various shades of blue. As Prospero spoke the closing lines of 5.1, he handed his spirit a vial of blue liquid:

> I'll deliver all,
> And promise you calm seas, auspicious gales,
> And sail so expeditious that shall catch
> Your royal fleet far off. *(aside to Ariel)* My Ariel, chick,
> That is thy charge. Then to the elements
> Be free, and fare thou well.
>
> (5.1.317–22)

After some hesitation, Ariel drank the blue liquid down. It spilled down his mouth and chest as he convulsed, before collapsing, seemingly dead, into Prospero's empty wheelchair. A tingle ran across my back and shoulders, just as the assembled crowd released the kind of communal gasp of pleasure one might more usually associate with firework displays. Because the moment Ariel had fallen, a cluster of blue, white, grey, and silver balloons had been released from behind the set, and surged upwards into the night sky. They continued to rise, diminishing but never quite disappearing from view, as Prospero delivered his final plea:

> Gentle breath of yours my sails
> Must fill, or else my project fails,
> Which was to please. Now I want
> Spirits to enforce, art to enchant;
> And my ending is despair
> Unless I be relieved by prayer,
> Which pierces so, that it assaults
> Mercy itself, and frees all faults.
> As you from crimes would pardoned be,
> Let your indulgence set me free.
>
> (Epilogue, 11–20)

Prayer, magic, release, transcendence: on the train home later, the analytical side of my mind would pick apart the experience I had just undergone. But at that moment, as the cast took their final bow, the audience cheered, and Ariel's liberated spirit continued to soar ever upwards on its journey towards heaven, I had been intoxicated, and the myth had been gloriously, seductively, and beautifully intact. It seems a shame to deconstruct it.

Notes

1   Popular Shakespeares

1. Readers might take a look at Graham Holderness's excellent *Cultural Shakespeare: Essays in the Shakespeare Myth* (2001) for insightful analyses of some of the various cultural uses to which the name 'Shakespeare' is put.
2. Would be, and indeed is. I direct the reader towards, for example, Levine (1991) and Bristol (1990, 1996).
3. This is certainly the argument put forward by Pierre Bourdieu in *Distinction: A Social Critique of the Judgement of Taste*. Here, Bourdieu distinguishes between the 'pure gaze' associated with high art, which looks in an abstract manner for the access the artwork provides to 'universal truths', and the 'popular aesthetic', which approaches the artwork in a manner no different from other areas of everyday life (1984: 4). In his attempt to define popular theatre, David Mayer similarly constructs a binary opposition between the 'aesthetic' and the 'popular' (1977: 257–77).
4. Alan Sinfield suggests that within two years, Peter Hall's desire to appeal to a 'popular' audience became 'refocused entirely in terms of the "the young" and particularly those in higher education' (Sinfield 2000: 179).
5. 'Talking Theatre', *The Tempest*: Mark Rylance, Shakespeare's Globe, 31 August 2005.
6. See Bennett (1990: 163–4); Elam (2005: 86–7); Ubersfeld (1981: 306).
7. S. L. Bethell (1977: 29) and Peter Davison (1982a: 1–11) put forward similar arguments.
8. Ronald Knowles makes a similar argument (1998: 36–60). For a more detailed discussion of Shakespeare and Carnival, see Bristol 1985.
9. In a book on stand-up comedians, Oliver Double writes of a similar 'personality spectrum' from 'character comedians' to the 'naked human being' (2005: 73–6). Double notes that 'the concept of a continuous spectrum from character to naked self, however, does not really capture the subtle interweaving of truth and fiction in the onstage identities of stand-up comedians' (2005: 77).
10. Indeed, Weimann himself has described Brecht as 'a fundamental inspiration' for his theory (Guntner et al. 1989: 231).
11. This could, it should be noted, be a misrepresentation of Brook. His comments on Shakespeare's plays frequently tread an ambiguous line between, on the one hand, claims for his universality, and on the other, suggestions of a potential for reinvention. His comments on *Hamlet* might be an example:

    > *Hamlet* is inexhaustible, limitless. Each decade brings with it new explanations, fresh interpretations. Yet *Hamlet* remains intact, a fascinating enigma. *Hamlet* is like a crystal ball, ever rotating. At each instant, it turns a new facet towards us and suddenly we seem to see the whole play more clearly.
    >
    > (Brook 2001)

2   Text and Metatext: Shakespeare and Anachronism

1. One might compare S. L. Bethell's analysis: 'Shakespeare, then, in the orthodox line of Tudor political philosophy, brings history into active relationship
   with contemporary life; he does not immerse himself in the past, but con-
   templates the past in the light of his own times' (1977: 51). On anachronism,
   Bethell comments: 'The co-presence of such contrasting elements renders
   doubly impossible any illusion of actuality; once again, the audience must
   necessarily remain critically alert, whilst at the same time the historical
   element gives current significance to an historical situation' (1977: 49).
2. This is not, it should be noted, to make a claim for a continuous tradition of
   anachronism in popular culture: anachronism in medieval and Renaissance
   drama was a product of that era's evolving concept of history, and made use
   of the tension between popular oral histories and official written ones. Rackin
   highlights the importance of the growing number of printed histories which
   emerged during the Renaissance – when histories were disseminated orally,
   anachronisms had been far more common (Rackin 1991: 234).
3. *Virgidemiarum* II, Liber I, Satire iii, 1597 (cited Gurr 2002: 219–20).
4. *The Defence of Poesie*, 1595 (*The Prose Works of Sir Philip Sidney*, ed. Albert
   Feuillerat, Cambridge, 1912, 3: 39).
5. Wilkinson, T. (1790) *Memoirs of His Own Life*, York, 4: 111 (cited Fisher 2003:
   63).
6. Self-described 'iambic fundamentalist' Peter Hall, for example, argues:

   > These are difficult times for the classical actor because there is little technical
   > consistency. I have worked in a theatre where the director before me urged
   > the actors to run on from one line to the next, speak the text like prose, and to
   > take breaths whenever they felt like it. He wanted them, he said, to be 'real'.
   > They were; but they weren't comprehensible.
   >
   > (2003: 11)

7. This story is related more fully in Maher 1992: 41. Conkie records that
   when Rylance's *Hamlet* asked the same questions at the Globe in 2000, he
   deliberately solicited audience responses, repeating the question until it was
   answered. He apparently received both negative and affirmative answers, altering his playing of the speech which followed accordingly (Conkie 2006: 38).
8. A different version of the same article also appeared in the Old Vic programme for *Richard II* under the heading, 'Shakespeare Our Contemporary'.
9. Wesker levels a similar charge at Trevor Nunn's 1999 production in an open
   letter on his website (Wesker 1999).

3   'A Play Extempore': Interpolation, Improvisation, and Unofficial Speech

1. See, for example, Bartlett's *Familiar Quotations* (1961, 13th edn., London:
   Macmillan) or *The Oxford Dictionary of Quotations* (1979, 3rd edn., OUP).
2. I. C. Media Productions, *The Wisdom of Shakespeare in As You Like It* (1998),
   The Wisdom of Shakespeare in The Merchant of Venice (1998), *The Wisdom of*

Shakespeare in Julius Caesar (1999), *The Wisdom of Shakespeare in The Tempest* (2000), and *The Wisdom of Shakespeare in Twelfth Night* (2002). For Dawkins' influence on Rylance's practice, see Peterson 2005.

3. This process is detailed in, for example, Taylor (1990), Levine (1991), Lanier (2002: 21–49), and Henderson (2007).

4. See K. M. Lea's discussion of this in *Italian Popular Comedy* (1934), New York: Russell & Russell, vol. II, 381.

5. The scenes featuring these characters are most accurately rendered; see Davies 2001: 179.

6. Wiles's version omits the line 'and, my coate wants a cullison', so I have reinstated it, following the pattern of Wiles's punctuation for the other lines.

7. Duthie objects to this theory on the grounds that 'allowing a Shakespearean cancellation here, he has nevertheless retained part of the attack upon the clown – he has retained the complaint about improvisation' (1941: 235).

8. To back up this point, Davison quotes Cleopatra, who feared that 'the quick comedians / Extemporally will stage us' (*Anthony and Cleopatra*, 5.2.212–3). As we saw in Chapter 2, however, the latter passage, much like the *Hamlet* speech, can be understood as a disjunctive anachronism: in other words, as a deliberate joke.

9. In a sequence reminiscent of Malvolio's gulling, Tarlton conveys one of his drunken fellow actors to a jail; upon waking, the drunkard is teased with moans 'that one so young should come to so shamefull a death as hanging' (Halliwell 1844: 31). Tarlton also deceives a madman who wishes to cut off his head (1844: 32); he gets out of paying an innkeeper at Sandwich by giving the impression that he is a Catholic priest (1844: 36–7). Very often, like Touchstone's escapades in the Forest of Arden, Tarlton's adventures take the form of the worldly entertainer from the city deceiving simple country folk: in 'How Tarlton frightened a country fellow', for example, Tarlton terrifies a 'simple country fellow in an alehouse' with an accusation of treason (Halliwell 1844:17); in another anecdote, he wriggles out of marrying a 'country wench' very reminiscent of *As You Like It*'s Audrey (1844: 33). Tarlton also becomes a Falstaff-like gull figure himself, being led into a trap, for example, by the promise of sex (1844: 39).

10. To support this claim, Rackin cites Sir Philip Sidney's remark in *The Defence of Poesy* that 'verse far exceeds prose in the knitting up of the memory' (1991: 238).

11. More official than it has any reason to be, one might argue, given the disputed authority of many of the texts.

12. Gary Taylor relates an anecdote of a scene improvised by comedians John Monteith and Suzanne Rand 'as if written by Shakespeare':

> The result was screamingly funny, but I did not hear a single quotation from Shakespeare; his style was suggested, instead, by acrobatic contortions of grammar, the occasional 'alas,' odd 'doth,' and frequent 'thee,' incongruous mixtures of orotund polysyllables and street slang, and a singsong approximation of blank verse.
>
> (1999: 203)

13. Actors from Campbell's continuing collaborators The Sticking Place formed *Shall We Shog?*'s London team. The Newcastle team are otherwise known as

The Suggestibles, an improvisation troupe who perform together regularly in their home town. The Liverpool team, meanwhile, were the winners of an event called *Farting Around in Disguises* which took place under Campbell's direction at the Liverpool Everyman in 2004; their team was called The Cottonwool Sandwiches. *Shall We Shog?* also featured much the same line-up of teams as *Clash of the Frightened*, a similar (though not Shakespeare-themed) event staged by Campbell in Liverpool in January 2005.

14. In an article for the *New Statesman* on the subject (which was, it should be noted, written following a radio discussion presented by Campbell), Michael Coveney quotes a 'nub' attributed to the early twentieth-century actor-manager Donald Wolfit:

> List, I sense a nubbing in far glens, where minnows swoop the pikey deep which is unpiked less pikey be, cross-bolted in their crispy muffs and choose the trammelled way. ... O freeze my soul in fitful sleep lest wind-filled sprites bequim the air and take us singly or in threes in mad agog or lumpsome nub, aghast to Milford Haven.
>
> (Coveney 2005)

15. When Brook writes about the Pompey and Barnadine scenes in *Measure for Measure*, he argues:

> To execute Shakespeare's intentions we must animate all this stretch of the play, not as fantasy, but as the roughest comedy we can make. We need complete freedom, rich improvisation, no holding back, no false respect – and at the same time we must take great care, for all round the popular scenes are great areas of the play that clumsiness can destroy.
>
> (1990: 99)

It is notable that Brook characterises this as a direct execution of 'Shakespeare's intentions'.

16. It is perhaps significant that like several other Globe performers, Lennox is also a stand-up comedian.

17. The line provoked a big laugh when I saw it; in a post-show talk, however, the actors noted that when Naiambana had tried this line the day before, it had been met with a frosty silence.

18. In the performance I saw, Garnon succeeded, with a particularly surreal line about Captain Birdseye and fish fingers. Rylance broke out of character, repeated the line very slowly, and after a second audience laugh, concluded, 'That's the silliest thing I've ever heard.'

4 'It's like a Shakespeare play!': Parodic Appropriations of Shakespeare

1. It should be noted that the plays of Shakespeare as they were performed during the Renaissance did not, according to Bakhtin, belong to such a 'lofty' and limited genre. Bakhtin makes it clear in his essays on parody and elsewhere that examples of parody's 'doubling effect' can be found quite clearly in the works of Shakespeare (1981: 79), and in this respect his analysis supports Robert Weimann's contention that Shakespeare's 'conjunction of two very different traditions' drew from popular drama in order to 'continually

undercut the pathos of literary representations by irreverently turning things around and upside down' (Guntner, Wekwerth & Weimann 1989: 233). Bakhtin and Weimann refer primarily to Shakespeare's use of Fool and Clown figures, but Shakespeare was not above sustained parody: the plays-within-plays of *A Midsummer Night's Dream* and *Love's Labour's Lost*, the archaic bombast of Pistol's speech, and arguably much of *As You Like It*'s pastoral lyricism, send up theatrical practices which had become outmoded in his own time. Today, however, Shakespeare can be (and, particularly in pop culture depictions, often is) flattened into just the sort of monologic, culturally authoritative genre which Bakhtin describes as the province of the 'lofty direct word'.

2. Rose, however, argues that 'parody may still be said to be "comic" even when its comic aspects are not noticed or understood by a recipient' (1993: 32).

3. See, for example, Schoch (2002), Wells (1965 and 1977), and Holland (2007).

4. Broadcast on BBC Radio 3 on 27 April 1992, with a cast including Peter Jeffrey, Harriet Walter, and Simon Russell Beale. Pontac's other Shakespearean parodies for Radio 4 include *Prince Lear* (1994) and *Fatal Loins* (2001), a *Romeo and Juliet* pastiche (Greenhalgh 2007: 194).

5. Barton's two-part television series *Word of Mouth* had been televised in 1980; his more famous series *Playing Shakespeare* would follow in 1984.

6. Holderness criticises the BBC Shakespeare series' 'remorselessly monumental classicising of both the plays and the concept of British culture into which they were assimilated' (2001: 13). The sketch might be seen as a similar criticism through parody.

7. Drakakis notes that Morecambe and Wise's *Antony and Cleopatra* parody similarly sends up Marlon Brando's performance in the Joseph Mankiewicz film of 1953 (1997: 168).

8. The episode is discussed in further detail in Lanier (2002: 106–7).

9. Robey was a notable Shakespeare enthusiast, claiming he kept a copy of *Hamlet* with him at all times (Harding 1990: 155), and in 1935 he made headlines when Sydney Carroll cast him as Falstaff in *1 Henry IV* at His Majesty's Theatre (he would later play the dying Falstaff in Olivier's film of *Henry V*). For more on Shakespeare and music hall, see Davison 1982a and 1982b.

10. Bergson felt it was the function of laughter to correct (by humiliation) social 'inelasticity'.

11. And, perhaps, in the way in which it broke its own rules. Most widely reported in reviews was the distinctly non-Shakespearean line, 'Thou wast merely commanded to blast the bloody portals off!'

12. 'Excellent', it should be noted for the uninitiated, is Mr Burns' catchword on *The Simpsons*.

13. Something similar has since been done on *The Simpsons* itself (though without as much direct incorporation of Shakespearean language): the 2002 episode 'Tales from the Public Domain' featured a retelling of *Hamlet* in which the play's roles were portrayed by *Simpsons* characters. Shakespeare's characters were adapted in order to overlap with the personae of the series: Marge's Gertrude, for example ('Hamlet, what'd I tell ya about running with swords?'), or Homer's Ghost ('Yes, I have returned from the dead!' 'Looks

like you've returned from the buffet ...'). When Shakespeare's language was appropriated, it was done in a more cynical manner:

HAMLET (BART).   Methinks the play's the thing,
Wherein I'll catch the conscience of the King!

CLAUDIUS (MOE).   'Catch my conscience?' What?

HAMLET.   You're not supposed to hear me! That's a soliloquy!

CLAUDIUS.   OK, well, I'll do a soliloquy too. *(clears throat)* 'Note to self: kill that kid.'

14. There is a strong historical precedent for such appropriation: illegitimate 'drolls' like Robert Cox's *The Merry Conceited Humours of Bottom the Weaver* (published in Francis Kirkman's anthology *The Wits: or, Sport upon Sport* in 1662) were performed in fairground booths and taverns during the Interregnum and afterwards, borrowing freely from Shakespeare's clowning scenes in order to construct shorter, popular entertainments.
15. The sketch was also broadcast on television on *The Music Box*, 1 February 1957.
16. A similar mockery of academia permeates the published script, with its many footnotes by the fictional 'Professor J. M. Winfield'.
17. A further example can be found in *Die Rundköpfe und die Spitzköpfe*. Isabella – the equivalent of her namesake in *Measure for Measure* – arranges for a working-class prostitute to replace her in Angelo's bed; the prostitute is subsequently beaten up. Though this does not directly follow the plot of *Measure for Measure*, it does perhaps (as Margot Heinemann suggests) expose the 'very nasty underworld of sexual and commercial exploitation of inferiors' which underpins Shakespeare's play: Mariana, trapped by her own poverty, is forced to stand in for Isabella in what is arguably a similar manner (Heinemann 1992: 220). For more on Brecht's Shakespearean adaptations, see Cohn 1976.
18. Such arguments owe much to studies such as, for example, Emrys Jones' *Scenic Form in Shakespeare* (1971). Here, Jones argues that Shakespeare creates a 'structure, an *occasion* – which may be said to be (however dangerous the phrase) independent of the words which are usually thought to give the scene its realisation' (1971: 3).
19. A similar debate was prompted on the same forum by Cheek by Jowl's international tour of *Twelfth Night*, which was performed by its Russian cast in their native language. Terence Hawkes articulated objections to both Synetic Theater and Cheek by Jowl's productions on the grounds that the cultural transcendence they implied for Shakespeare was in fact a form of Anglophone cultural imperialism (Hawkes 2007; Galbi, Manger, Drakakis et al. 2007).
20. Peter, as the reader might recall from Chapter 2, made a similar analysis of Vesturport's *Romeo and Juliet*, describing it as 'a spoof, a romp, a game, even a loving homage. It is hugely enjoyable, but one thing it isn't is Shakespeare' (*Sunday Times*, 12 October 2003).
21. The production was broadcast on 24 June 1987, on the PBS show *Live From Lincoln Center*, and my transcriptions are drawn from a recording of this transmission.

22. The troupe's very name, in fact – referencing and transforming the title of a novel by Dostoyevsky – indicates a decidedly carnivalesque attitude towards the established literary canon.

23. Indeed, the irreverent, circus-themed *Romeo and Juliet* by Vesturport discussed both in Chapter 2 and in the Personal Narrative strand might be seen as this production's most direct successor.

24. The same description has been frequently applied to Baz Luhrmann's film *Romeo + Juliet* and various other 'Shakespop' films (such as those mentioned above), and has also been used in press discussions of the Reduced Shakespeare Company, BBC3's *From Bard to Verse*, and Propeller's production of *Rose Rage*, among others. Toby Young, who wrote *The Spectator*'s review of *Bomb-itty*, has used the analogy before himself, in his review of the New York Shakespeare Festival's 1995 production of *The Tempest*.

5 Shakespeare's Popular Audience: Reconstructions and Deconstructions

1. Brown articulates something similar in *New Sites For Shakespeare* (1999: 95).

2. 'Talking Theatre', *Romeo and Juliet*: Callum Coates (Paris) and Simon Müller (Tybalt), Shakespeare's Globe, 11 August 2004. During this talk, Coates and Müller revealed that Tim Carroll had directed the play without an explicit 'concept', and that apart from entrances and exits, fights, crowd scenes, and dances, blocking was not fixed: the playing of a scene, they said, would change from performance to performance. In early rehearsals, apparently, they had followed the 'Original Shakespeare Company' policy of working from Cue Scripts (Coates himself has acted with the OSC).

3. Bharucha's discussion is of Brook's *Mahabharata*, but similar criticisms apply to most of Brook's cultural borrowings from India.

4. I borrow this word from Graham Holderness. Writing in 1992, Holderness suggested that such mythologising tendencies – Rylance's interests apparently included Rosicrucianism, Freemasonry, ley lines, the phases of the moon, and sacred sites – implied that the Globe was, for the time being, unlikely to provide 'an unillusioned grasp of history' (2001: 102, 103).

5. In a presentation on the Elizabethan public playhouses by first-year undergraduates at the University of Kent, one student pointed out, as an aside, 'where the Queen would have sat'. When I questioned her as to her source for this dubious information, she rather sheepishly admitted it was *Shakespeare in Love*.

6. *Shakespeare in Love* does not, it should be noted, depict the Globe, but rather the Rose and the Curtain theatres.

7. The ideological implications of the sequence are even clearer in Davies' description:

> The Globe sequence covers a broad range of society as audience, from aristocrats to rowdy groundlings. ... There is a vigorous if rough-hewn camaraderie about the totality of the theatrical experience which mirrors that required of troops and generals in a war.
>
> (2000: 168–9)

8. Indeed, Madden indicates an allegiance to a 'shared experience' style theatre on the commentary track:

   > Played in broad daylight, the audience could see every other member of the audience – could often see the actors waiting to come onstage – but the active imagination they brought to it to create the world that was being described makes their belief even more powerful.

9. A through line of references to J. K. Rowling's *Harry Potter* series also plays with the postmodern collapsing of barriers separating elite and popular literary forms: Shakespeare eventually defeats the Carrionites with the *Potter* spell 'Expelliarmus!'.

10. In addition to the films by Branagh and O'Haver discussed here, Burnett considers James Callis and Nick Cohen's *Beginner's Luck* (2001), and Roger Goldby's *Indian Dream* (2003), which revolve around performances of *The Tempest* and *A Midsummer Night's Dream* respectively.

11. *Dead Poets Society*, in fact, makes use of similar standing ovations elsewhere in the film.

12. I am indebted to Miles Gregory for his analysis of this sequence in his unpublished conference paper, '"A Tale Told by a Moron": Shakespearean appropriation and cultural politics in Kevin Costner's *The Postman*'.

13. Willis notes, however, that the theme park's potentially progressive 'activation' of the passive spectator can be (as in the case of Disneyland) 'foreclosed by the way the amusement park is not conceived as a site of production but is felt instead to be a commodity itself' (1991: 16). For a more critical account of the 'theme park' aspects of Globe spectatorship, see Henderson (2002).

14. Semiotic analysis of theatre tends to focus on the signifying process that goes on between actor and spectator; the signifying channels between spectator and spectator are often neglected. Elam, for example, touches upon this only in one paragraph of *The Semiotics of Theatre and Drama* (2005: 87).

6   Shakespeare, Space, and the 'Popular'

1. Bharata's *Natya Shastra* states that theatres should be no bigger than 64 hastas long and 32 hastas wide; Edström tells us that 32 hastas is slightly less than 20 metres (1990: 11). 66 feet (20.1m) was considered 'intimate' by the designers of Chichester Festival Theatre ('the length of a cricket pitch', points out Mackintosh), while Laurence Olivier felt that 65 feet (19.8m) was the maximum acceptable distance between audience and performer (Mackintosh 1993: 106). Cheek By Jowl's designer Nick Ormerod suspects that the 'magic figure' is 21m (68.9 feet; Mulryne & Shewring 1995: 104), while Pauline Kiernan points out that the furthest distance between a member of the audience and the centre of the stage at the new Globe is about 50 feet (15.2m; Kiernan 1999: 19). It may be worth mentioning that the Driving Standards Agency specifies 20m (65.6 feet) as the minimum distance at which drivers should be able to make out number plates.

2. For some detailed accounts of the Elizabethan stage, see Foakes (1985), Gurr (1980), Hodges (1968), or, of course, Chambers (1923). For a brief overview, see Hosley (1971).

3. Indeed, Sir Alexander Wengrave's first long speech in Middleton's *The Roaring Girl* (from which the quotation on the previous page was taken) is surely another example of such delivery (1.1.131–53).
4. See Schoch (2002), Wells (1965), Holland (2007), and Lanier (2002: 21–49).
5. Of course, some will disagree. Mackintosh, a great champion of Victorian theatres, argues that the texts enable interaction with the upper galleries in the manner of a music-hall comedian, since 'Shakespeare offers opportunities for asides to the underprivileged wherever they are sitting'. He cites Ian McKellen's performance as Richard III in the National Theatre's famous production as one which owed a debt to McKellen's childhood experiences of variety theatre (1993: 136).
6. It was only after the proliferation of the mass media that 'legitimate' theatre, no longer the most cost-effective means of distributing mimetic art, began to reclaim a counter-cultural position and to adopt formations which recaptured something of the lost thrust stage. It is significant that all but a few of the most emphatically 'commercial' theatre productions of recent years have retained the picture-frame stage.
7. The productions and issues raised in this paragraph are discussed in further detail in my article 'A Shared Experience: Shakespeare and Popular Theatre' (2005).
8. After Joseph's death in 1967, the theatre was taken over by Alan Ayckbourn and moved to new premises in the former Scarborough Boys' High School in 1976, and then the former Odeon cinema building in 1988.
9. The number of Shakespearean plays (and indeed Renaissance plays in general) featuring swordfights or other physical scuffles is quite striking; I wonder whether the phenomenological effects of proximity to physical danger are today a largely neglected aspect of the way in which we understand the plays to work. It was not unknown, after all, for Elizabethan playgoers to be quite seriously harmed during the course of a play. Mark Rylance describes the 'physical activeness of the body whether standing or seated on a bench at the Globe' as a 'quite different state for the heart and mind', leading to 'an awakened and sometimes drenched sense of the physical body' (Kiernan 1999: 132).
10. The Globe experience is not as fully 'authentic' as it could be in its actor/audience relationship, either – the stage door is distinctly modern and removed from sight.
11. And indeed such a claim could also be made for 'natural' open-air spaces. 'Nature' itself, suggests Lefebvre, 'as apprehended in social life by the sense organs, has been modified and therefore in a sense produced' (1991: 68). Aldous Huxley's novel *Brave New World* gives an indication as to the ways in which nature might be 'produced' differently in an altogether different society.

7 Shakespearean 'Samples'

1. Shaughnessy describes the performance in detail on pages 189–93 (2002).
2. *Hot Fuzz*, it should be noted, is itself a pastiche of the action film genre, filled with intertextual filmic references.

Bibliography

Books, articles and interviews

Adams, J. Q. (1917) (ed.) *The Dramatic Records of Sir Henry Herbert*, New Haven: Yale University Press.

Addenbrooke, D. (1974) *The Royal Shakespeare Company: The Peter Hall Years*, London: William Kimber.

Alfreds, M. (1979) *A Shared Experience: The Actor as Story-teller*, Dartington: Department of Theatre, Dartington College of Arts.

Alfreds, M. & Barker, C. (1981) 'Shared Experience: from Science Fiction to Shakespeare', *Theatre Quarterly*, vol. x, no. 39, 12–22.

Allen, T. (2002) *Attitude: The Secret of Stand-Up Comedy*, Glastonbury: Gothic Image Publications.

Anderegg, M. (2003) 'James Dean meets the pirate's daughter: Passion and parody in *William Shakespeare's Romeo + Juliet* and *Shakespeare in Love*', in Burt & Boose 2003: 56–71.

Armitstead, C. (1994) 'The Trial of Shylock', *The Guardian*, 13 April, 4.

Armstrong, G. (1999) *Shylock*, London: The Players' Account.

Auden, W. H. (1968) *The Dyer's Hand & Other Essays*, New York: Vintage.

Auslander, P. (1997) *From Acting to Performance*, London: Routledge.

Auslander, P. (1999) *Liveness: Performance in a Mediatized Culture*, London: Routledge.

Bakhtin, M. (1981) *The Dialogic Imagination*, trans. Caryl Emerson, Michael Holquist, ed. Michael Holquist, Austin: University of Texas Press.

Bakhtin, M. (1984) *Rabelais and His World*, trans. H. Iswolsky, Bloomington: Indiana University Press.

Barish, J. (1996) 'Hats, Clocks and Doublets: Some Shakespearean Anachronisms' in J. M. Mucciolo (ed.) *Shakespeare's Universe: Renaissance Ideas and Conventions*, Aldershot: Scolar, 29–36.

Barthes, R. (1974) *S/Z: An Essay*, trans. Richard Miller, New York: Hill & Wang.

Barthes, R. (1977) 'The Death of the Author', *Image-Music-Text*, London: Fontana.

Baudrillard, J. (1988) *America*, trans. C. Turner, London: Verso.

Bayne, R. (1910) 'Lesser Elizabethan Dramatists' in A. W. Ward & A. R. Waller (eds) *The Cambridge History of English Literature Vol. 5: The Drama To 1642, Part One*, New York: G. P. Putnam's Sons, <http://www.bartleby.com/215/1303.html>.

Beckermann, B. (1970) *Dynamics of Drama*, New York: Alfred A. Knopf.

Bennett, A., Cook, P., Miller, J. & Moore, D. (2003) *The Complete Beyond the Fringe*, London: Methuen.

Bennett, S. (1990) *Theatre Audiences: A Theory of Production and Reception*, London: Routledge.

Berger, H. (1998) 'The Prince's Dog: Falstaff and the Perils of Speech-Prefixity', *Shakespeare Quarterly* 49, 40–73.

Bergson, H. (1956) 'Laughter', in Sypher, W. (ed.) *Comedy*, New York: Doubleday Anchor, 59–190.

Bethell, S. L. (1977) *Shakespeare and the Popular Dramatic Tradition*, New York: Octagon.

Bharucha, R. (1993) *Theatre and the World*, London: Routledge.

Blau, H. (1990) *The Audience*, Baltimore: Johns Hopkins University Press.

Boal, A. (2000) *Theater of the Oppressed*, trans. Charles A. & Maria-Odilia Leal McBride & Emily Fryer, London: Pluto Press.

Bolton, M. J. (2007) review of *Hamlet*, *Shakespeare Bulletin* 25.3, 83–6.

Borgeson, J., Long, A. & Singer, D. (1994) *The Compleat Works of Wllm Shkspr (abridged)*, New York: Applause Books.

Bourdieu, P. (1984) *Distinction: A Social Critique of the Judgement of Taste*, Cambridge, MA: Harvard UP.

Branagh, K. (1995) *In The Bleak Midwinter*, London: Nick Hern Books.

Brecht, B. (1965) *The Messingkauf Dialogues*, trans. J. Willett, Chatham: W. & J. Mackay & Co.

Brecht, B. (1967) 'Estranging Shakespeare', *The Drama Review*, 12.1, 108–11.

Brecht, B. (1973) *Collected Plays 9*, trans. Manheim, R & Sauerländer, W London: Methuen.

Brecht, B. (1977) *Brecht on Theatre*, trans. J. Willett, London: Eyre Methuen.

Brecht, B. (1979) *Collected Plays 2: 2*, trans. R. Manheim & J. Willett, London: Methuen.

Brecht, B. (1990) *The Measures Taken and Other Lehrstücke*, trans. R. Manheim & J. Willett, London: Methuen.

Bristol, M. D. (1985) *Carnival and Theatre: Plebian Culture and the Structure of Authority in Renaissance England*, London: Methuen.

Bristol, M. D. (1990) *Shakespeare's America/America's Shakespeare*, London: Routledge.

Bristol, M. D. (1996) *Big-Time Shakespeare*, London: Routledge.

Brockes, E. (2006) 'Fail again, fail better', *Guardian*, G2, 28 June, 18.

Brook, P. (1978) 'Lettre à une Etudiante Anglaise', in *Timon d'Athènes*, Paris.

Brook, P. (1989) *The Shifting Point: Forty years of theatrical exploration 1946–87*, London: Methuen.

Brook, P. (1990) *The Empty Space*, London: Penguin.

Brook, P. (2001) Introduction to *The Tragedy of Hamlet* programme, London: Young Vic Theatre.

Brown, J. R. (1974) *Free Shakespeare*, London: Heinemann.

Brown, J. R. (1993) *Shakespeare's Plays in Performance*, New York: Applause Books (revised edition).

Brown, J. R. (1999) *New Sites for Shakespeare: Theatre, the Audience and Asia*, London/New York: Routledge.

Brown, J. R. (2004) interviewed by Steve Purcell in London, 3 December.

Bryant, J. A., Jr. (1986) *Shakespeare and the Uses of Comedy*, Lexington: University of Kentucky Press.

Burnett, M. T. (2007) *Filming Shakespeare in the Global Marketplace*, Basingstoke: Palgrave Macmillan.

Burt, R. (2002) (ed.) *Shakespeare After Mass Media*, New York: Palgrave Macmillan.

Burt, R. (2003) 'Shakespeare, "Glo-cali-zation," race, and the small screens of post-popular culture', in Burt & Boose 2003: 14–36.

Burt, R. & Boose, L. E. (2003) (eds) *Shakespeare, the Movie, II: Popularizing the Plays on Film, TV, video, and DVD*, London and New York: Routledge.

Carlson, M. (1996) *Performance: A Critical Introduction*, London: Routledge.

Cartelli, T. (2003) 'Shakespeare and the street: Pacino's *Looking for Richard*, Bedford's *Street King*, and the common understanding', in Burt & Boose 2003: 186–99.

Castledine, A. (2006) interviewed by Steve Purcell in Eastbourne, 14 May.

Chambers, E. K. (1923) *The Elizabethan Stage*, Oxford: Clarendon Press.

Chambers, E. K. (1930) *William Shakespeare: A Study of Facts and Problems*, vol. 1, Oxford: Clarendon Press.

Cheek by Jowl (2007) *Cymbeline Script – Draft 7* (unpublished adaptation).

Cohn, R. (1976) *Modern Shakespeare Offshoots*, Princeton: Princeton University Press.

Conkie, R. (2006) *The Globe Theatre Project: Shakespeare and Authenticity*, New York: The Edwin Mellen Press.

Conkie, R. & Cuming, R. (2003) 'Circo Shakespeare', *Total Theatre Magazine*, vol. 15, 1, 6–8.

Cook, W. (2001) *The Comedy Store: The Club that Changed British Comedy*, London: Little, Brown.

Cousin, G. (1993) 'Footsbarn: From a Tribal "Macbeth" to an Intercultural "Dream"', *New Theatre Quarterly*, vol. IX, no. 33, 16–30.

Counsell, C. (1996) *Signs of Performance: An Introduction to Twentieth-Century Theatre*, London: Routledge.

Coveney, M. (2005) 'Aye, there's the nub', *New Statesman*, 28 March.

Creation Theatre (2006) 'Creation Theatre Company' webpage, <http://www.creationtheatre.co.uk/>. Accessed 4 February 2006.

Davies, A. (2000) 'The Shakespeare Films of Laurence Olivier' in Jackson 2000: 163–82.

Davies, A. (2001) 'Hamlet, Prince of Denmark' in M. Dobson & S. Wells (eds) *The Oxford Companion to Shakespeare*, Oxford: Oxford University Press.

Davison, P. (1982a) *Popular Appeal in English Drama to 1850*, London: Macmillan.

Davison, P. (1982b) *Contemporary Drama and the Popular Dramatic Tradition in England*, London: Macmillan.

Dent, R. W. (1981) *Shakespeare's Proverbial Language*, Berkeley: University of California Press.

Dentith, S. (2000) *Parody*, London & New York: Routledge.

Desmet, C. & Sawyer, R. (1999) (eds) *Shakespeare and Appropriation*, London and New York: Routledge.

Dobson, M. (2002) 'Shakespeare Performances in England, 2001' in *Shakespeare Survey* 55, 285–321.

Dobson, M. (2004) 'Shakespeare Performances in England, 2003' in *Shakespeare Survey* 57, 258–89.

Dobson, M. (2006) 'Shakespeare Performances in England, 2005' in *Shakespeare Survey* 59, 298–337.

Dobson, M. (2007) 'Shakespeare Performances in England, 2006' in *Shakespeare Survey* 60, 284–319.

Dobson, M. & Wells, S. (2001) (eds) *The Oxford Companion to Shakespeare*, Oxford: Oxford University Press.

Dolan, J. (2001) 'Performance, Utopia, and the "Utopian Performative"', *Theatre Journal* 53.3, 455–79.

Dollimore, J. & Sinfield, A. (1992) (eds) *Political Shakespeare: Essays in Cultural Materialism*, Manchester: Manchester University Press.

Double, O. (1997) *Stand-Up!: On Being a Comedian*, London: Methuen.

Double, O. (2005) *Getting The Joke: The Inner Workings of Stand-Up Comedy*, London: Methuen.

Drakakis, J. (1997) 'Shakespeare in Quotations' in S. Bassnett (ed.) *Studying British Cultures: An Introduction*, London & New York: Routledge, 156–76.

Duthie, G. I. (1941) *The 'Bad' Quarto of Hamlet: A Critical Study*, Cambridge: Cambridge University Press.

Eagleton, T. (1988) 'Afterword' in G. Holderness (ed.) *The Shakespeare Myth*, Manchester: Manchester University Press.

Edgar, D. (1988) *The Second Time As Farce: Reflections on the Drama of Mean Times*, London: Lawrence & Wishart.

Edgar, D. (1999) *State of Play: Playwrights on Playwriting*, London: Faber.

Edström, P. (1990) *Why Not Theaters Made For People?*, trans. C. Forslund, Värmdö: Arena Theatre Institute Foundation.

Elam, K. (2005) *The Semiotics of Theatre and Drama*, London: Methuen.

Elliott, M. (1973) 'On Not Building For Posterity', Mulryne & Shewring 1995: 16–20.

Eyre, R. (1993) *Utopia and Other Places*, London: Bloomsbury.

Farrell, J. (1991) 'Introduction' to Fo, D. *Tricks of the Trade*, trans. J. Farrell, London: Methuen.

Fenton, D. (1930) *The Extra-Dramatic Moment in Elizabethan Plays before 1616*, Philadelphia: University of Pennsylvania.

Fisher, J. W. (2003) 'Audience Participation in the Eighteenth-Century London Theatre' in S. Kattwinkel (ed.) *Audience Participation: Essays on Inclusion in Performance*, Westport, CT: Praeger, 55–69.

Fisher, M. (2007) 'Juxtaposition and synthesis', in Edinburgh International Festival, *La Didone* programme, Edinburgh: Royal Lyceum Theatre.

Foakes, R. A. (1985) *Illustrations of the English Stage, 1580–1642*, London: Scolar Press.

Folkerth, W. (2000) 'Roll Over Shakespeare: Bardolatry Meets Beatlemania in the Spring of 1964', *Journal of American Culture* 23 (4), 75–80.

Freeman, J. (2003) 'Shakespeare's Rhetorical Riffs' in T. J. McGee (ed.) *Improvisation in the arts of the Middle Ages and Renaissance*, Kalamazoo, MI: Medieval Institute Publications, 247–72.

Freud, S. (1976) *Jokes and their Relation to the Unconscious*, trans. J. Strachey, ed. J. Strachey & A. Richards, London: Pelican, 132–61.

Furness, H. H. (1877) (ed.) *Hamlet: A New Variorum Edition of Shakespeare*, New York: Dover Publications.

Galbi, D., Manger, S., Drakakis, J. and others, 'Wordless Macbeth', online postings, 16–29 January 2007, *SHAKSPER: The Global Electronic Shakespeare Conference.* 4 August 2007 <http://www.shaksper.net/archives/2007/0028.html>.

Garnon, J. (2006) interviewed by Steve Purcell in Bristol, 6 March.

Genette, G. (1982) *Palimpsestes*, Paris: Seuil.

Granville-Barker, H. (1922) *The Exemplary Theatre*, London: Chatto & Windus.

Greenhalgh, S. (2007) 'Shakespeare Overheard: Performances, Adaptations, and Citations on Radio', in Shaughnessy 2007: 175–98.

Gregory, M. (2007) '"A Tale Told by a Moron": Shakespearean appropriation and cultural politics in Kevin Costner's *The Postman*' (unpublished paper).

Griffiths, H. (2004) 'The Geographies of Shakespeare's Cymbeline', *English Literary Renaissance* 34.3, 359–86.

Grotowski, J. (1969) 'Towards a Poor Theatre' in E. Barba (ed.) *Jerzy Grotowski: Towards a Poor Theatre*, London: Methuen.

Guntner, J. L., Wekwerth, M. & Weimann, R. (1989) 'Manfred Wekwerth and Robert Weimann: "Brecht and Beyond"' in J. L. Guntner & A. M. McLean, (1998) (eds) *Redefining Shakespeare: Literary Theory and Theater Practice in the German Democratic Republic*, London: Associated University Press, 226–40.

Gurr, A. (1980) *The Shakespearean Stage, 1574–1642* (2nd edn.), Cambridge: Cambridge University Press.

Gurr, A. (2002) *Playgoing in Shakespeare's London* (2nd edn.), Cambridge: Cambridge University Press.

Guthrie, T. (1961) *A Life in the Theatre*, London: Readers Union.

Hall, P. (2003) *Shakespeare's Advice to the Players*, London: Oberon Books.

Hall, P. (2004) 'My Dream Come True', *The Independent*, 2 December.

Hall, S. (1998) 'Notes on deconstructing "the Popular"' in J. Storey (ed.) *Cultural Theory and Popular Culture: A Reader* (second edition), Hemel Hempstead: Prentice Hall, 442–53.

Halliwell, J. O. (1844) (ed.) *Tarlton's Jests, and News Out of Purgatory*, London: The Shakespeare Society.

Harbage, A. (1969) *Shakespeare's Audience*, New York: Columbia University Press.

Harding, J. (1990) *George Robey and the Music Hall*, Sevenoaks: Hodder & Stoughton.

Hartley, A. J. (2005) *The Shakespearean Dramaturg: A Theoretical and Practical Guide*, New York: Palgrave Macmillan.

Hawkes, T. (1969) (ed.) *Coleridge on Shakespeare*, Harmondsworth: Penguin.

Hawkes, T. (1986) *That Shakespeherian Rag*, London & New York: Methuen.

Hawkes, T. (1992) *Meaning by Shakespeare*, London: Routledge.

Hawkes, T. (1996) (ed.) *Alternative Shakespeares 2*, London: Routledge.

Hawkes, T. (2002) *Shakespeare in the Present*, London: Routledge.

Hawkes, T. (2007) 'Russian Twelfth Night', online posting, 23 November 2006, *SHAKSPER: The Global Electronic Shakespeare Conference*. 4 August 2007 <http://www.shaksper.net/archives/2006/1055.html>.

Heinemann, M. (1992) 'How Brecht read Shakespeare' in Dollimore & Sinfield 1992: 202–30.

Henderson, D. E. (2002) 'Shakespeare the Theme Park', in Richard Burt (ed.) *Shakespeare After Mass Media*, New York: Palgrave Macmillan, 109–18.

Henderson, D. E. (2007) 'From Popular Entertainment to Literature' in Shaughnessy 2007: 6–25.

Hewison, R. (1995) 'The Empty Space and the Social Space: A View From the Stalls', Mulryne & Shewring 1995: 52–60.

Hodges, C. W. (1968) *The Globe Restored: A Study of the Elizabethan Theatre* (2nd edn.), Oxford: Oxford University Press.

Holderness, G. (1988) (ed.) *The Shakespeare Myth*, Manchester: Manchester University Press.

Holderness, G. (2001) *Cultural Shakespeare: Essays in the Shakespeare Myth*, Hatfield: University of Hertfordshire Press.

Holland, P. (1997) *English Shakespeares: Shakespeare on the English Stage on the 1990s*, Cambridge: Cambridge University Press.

Holland, P. (2007) 'Shakespeare abbreviated' in Shaughnessy 2007: 26–45.

Holmberg, A. (1983) review of *The Comedy of Errors, Performing Arts Journal*, vol. 7, no. 2, 52–5.

Homan, S. (1989) *The Audience as Actor and Character*, Lewisburg: Bucknell University Press.

Hosley, R. (1971) 'The Playhouses and the Stage' in K. Muir & S. Schoenbaum (eds) *A New Companion To Shakespeare*, Cambridge: Cambridge University Press, 15–34.

Howerd, F. (1977) *On the Way I Lost It*, London: Star.

Hutcheon, L. (1985) *A Theory of Parody: The Teachings of Twentieth-Century Art Forms*, New York & London: Methuen.

Huxley, A. (2004) *Brave New World*, London: Vintage.

iO Theater (2007) 'The Improvised Shakespeare Company' webpage, <http://www.iochicago.net/s_shakespeare.php>. Accessed 9 September 2007.

Jackson, R. (1996) 'Rehearsal to Wrap' in Branagh, K. *Hamlet: Screenplay, Introduction and Film Diary*, London: Chatto & Windus.

Jackson, R. (2000) (ed.) *The Cambridge Companion to Shakespeare on Film*, Cambridge: Cambridge University Press.

Jiwani, S. (2005) 'India is an act of theatre', *DNA Sunday*, 5 November.

Johnson, S. (1965) *Johnson on Shakespeare* (ed.) W. Raleigh, Oxford: Oxford University Press.

Johnstone, K. (1981) *Impro: Improvisation and the Theatre*, London: Eyre Methuen.

Johnstone, K. (1999) *Impro for Storytellers*, London: Faber & Faber.

Jones, E. (1971) *Scenic Form in Shakespeare*, Oxford: Clarendon Press.

Joseph, S. (1968) *New Theatre Forms*, New York: Theatre Arts Books.

Kattwinkel, S. (2003) (ed.) *Audience Participation: Essays on Inclusion in Performance*, Westport, CT: Praeger.

Katz, M. (2004) *Capturing Sound: How Technology Has Changed Music*, Berkeley: University of California Press.

Kaye, N. (1996) *Art Into Theatre: Performance Interviews and Documents*, Amsterdam: Harwood Academic Publishers.

Kaye, N. (2000) *Site Specific Art: Performance, Place and Documentation*, London: Routledge.

Kershaw, B. (1992) *The Politics of Performance: Radical theatre as cultural intervention*, London: Routledge.

Kershaw, B. (1999) *The Radical in Performance: Between Brecht and Baudrillard*, London: Routledge.

Kiernan, P. (1999) *Staging Shakespeare at the New Globe*, Basingstoke: Palgrave Macmillan.

Klotz, H (1988) *The History of Postmodern Architecture*, Cambridge, MA: MIT Press.

Knowles, R. (1998) 'Carnival and Death in *Romeo and Juliet*' in R. Knowles (ed.) *Shakespeare and Carnival: After Bakhtin*, Basingstoke: Palgrave Macmillian, 36–60.

Koestler, A. (1976) *The Act of Creation*, London: Picador.

Lamb, C. (1963) 'On the Tragedies of Shakespeare, Considered with Reference to their Fitness for Stage Representation', in *Charles Lamb: Essays* (ed.) R. Vallance and J. Hampden, London: Folio Society, 20–37.

Lan, D. (2005) Interview with David Lan in the *As You Like It* programme, London: Wyndham's Theatre.

Lanier, D. (2002) *Shakespeare and Modern Popular Culture*, Oxford: Oxford University Press.

Lanier, D. (2003) 'Nostalgia and theatricality: The fate of the Shakespearean stage in the *Midsummer Night's Dreams* of Hoffman, Noble, and Edzard', in Burt & Boose 2003: 154–72.

Lea, K. M. (1934) *Italian Popular Comedy*, New York: Russell & Russell, vol. ii.

Lefebvre, H. (1991) *The Production of Space*, trans. Donald Nicholson-Smith, Oxford: Basil Blackwell.

Levine, L. W. (1991) 'William Shakespeare and the American People: A Study in Cultural Transformation' in Chandra Mukerji and Michael Schudson (eds) *Rethinking Popular Culture*, Berkeley: University of California Press.

Longhurst, D. (1988) '"You Base Football Player!": Shakespeare and Contemporary Popular Culture' in Graham Holderness (ed.) *The Shakespeare Myth*, Manchester: Manchester University Press.

Luscombe, C. & McKee, M. (1994) (eds) *The Shakespeare Revue*, London: Nick Hern Books.

McAuley, G. (2000) *Space in Performance: Making Meaning in the Theatre*, Ann Arbor: University of Michigan Press.

McConachie, B. A. (1998) 'Approaching the "Structure of Feeling" in Grassroots Theatre', *Theatre Topics* 8: 1, March, 33–53.

McGrath, J. (1974) *The Cheviot, the Stag, and the Black, Black Oil*, London: Methuen.

McGrath, J. (1996) *A Good Night Out: Popular Theatre: Audience, Class and Form*, London: Nick Hern Books

Mackintosh, I. (1993) *Architecture, Actor & Audience*, London: Routledge.

MacHomer (2007) 'Class bookings' webpage, <http://www.machomer.com>. Accessed 1 August 2007.

Maher, M. Z (1992) *Modern Hamlets and their Soliloquies*, University of Iowa Press.

Marowitz, C. (1969) *The Marowitz Hamlet & The Tragical History of Doctor Fautsus*, London: Penguin.

Marowitz, C. (1991) *Recycling Shakespeare*, London: Macmillan.

Mason, B. (1992) *Street Theatre and Other Outdoor Performance*, London: Routledge.

Mayer, D. (1977) 'Towards a Definition of Popular Theatre' in D. Mayer & K. Richards (eds) *Western Popular Theatre*, London: Methuen, 257–77.

Miller, J. (1988) 'Jokes and Joking: A Serious Laughing Matter', in J. Durant & J. Miller (eds) *Laughing Matters*, Harlow: Longman, 5–16.

Mitter, S. (1992) *Systems of Rehearsal: Stanislavsky, Brecht, Grotowski and Brook*, London: Routledge.

Mulryne, R. & Shewring, M. (1995) (eds) *Making Space for Theatre: British Architecture and Theatre Since 1958*, Stratford-upon-Avon: Mulryne & Shewring.

Mulryne, R. & Shewring, M. (1997) (eds) *Shakespeare's Globe Rebuilt*, Cambridge: Cambridge University Press.

Nabokov, V. (1973) *Strong Opinions*, New York: McGraw-Hill.

Naughton, J. (2006) 'Hit or myth', *Radio Times*, 7–13 October, 15.

Nicholson, B. (1880) 'Kemp and the Play of *Hamlet* – Yorick and Tarlton – A Short Chapter in Dramatic History', *The New Shakespeare Society's Transactions (1880-2)*, London: Trübner and Ludgate Hill, Part 1, 57–66.

Norman, M. & Stoppard, T. (1998) *Shakespeare in Love: A Screenplay*, New York: Hyperion/Miramax Books.

Northern Broadsides (2006) 'About us' webpage, <http://www.northern-broadsides. co.uk/PAGES/aboutus.htm>. Accessed 4 February 2006.

Nunn, T. (2005) 'Play for Today', *The Guardian*, 24 September, 14.

Old Vic (2005) *Richard II* seating plan, <http://www.oldvictheatre.com>. Accessed 26 November 2005.

Olivier, L. (1986) *On Acting*, London: Weidenfeld & Nicolson.

Oswald, P. (2005) *The Storm, or 'The Howler': An Appalling Mistranslation by Peter Oswald of a Roman Comedy by Plautus*, London: Oberon Books.

Out of Joint (2005) 'International & UK tour for Africa-infused *Macbeth*', press release, 26 August, <http://www.outofjoint.co.uk/prods/MAC_ PRESSRELEASE05.doc>. Accessed 26 November 2005.

Pennington, M. (2004) *Hamlet: A User's Guide*, London: Nick Hern Books.

Peterson, S. (2005) 'Art and Soul', *Caduceus* 66, Spring, 6–9.

Pitcher, J. (2005) (ed.) *Cymbeline*, London: Penguin.

Purcell, S. (2005) 'A Shared Experience: Shakespeare and popular theatre' in W. Sharman & P. Holland (eds) *Performance Research: On Shakespeare*, Routledge Journals, vol. 10, no. 3, 74–84.

Rackin, P. (1991) *Stages of History: Shakespeare's English Chronicles*, London: Routledge.

Reduced Shakespeare Company (2007) 'The Complete Works of William Shakespeare (abridged)' webpage, <http://www.reducedshakespeare.com/ shakespeare.php>. Accessed 9 September 2007.

Rice, E. & Grose, C. (2007) *Cymbeline*, London: Oberon Books.

Roach, J. (1996) *Cities of the Dead: Circum-Atlantic Performance*, New York: Columbia University Press.

Rose, M. (1979) *Parody/Metafiction: An Analysis of Parody as a Critical Mirror to the Writing and Reception of Fiction*, London: Croom Helm.

Rose, M. (1993) *Parody: Ancient, Modern, and Post-Modern*, Cambridge: Cambridge University Press.

RSC (2004) 'RSC outlines plan for renewal of theatre', press release, 22 September, <http://www.rsc.org.uk/press/420_1700.aspx>. Accessed 26 November 2005.

RSC (2005) *Julius Caesar* and *The Two Gentlemen of Verona* publicity material.

Rude Mechanical Theatre Co. (2006) 'About the Company' webpage, <http://www. therudemechanicaltheatre.co.uk/about_the_co.html>. Accessed 4 February 2006.

Rutter, C. C. (2007) 'Shakespeare's Popular Face: From the Playbill to the Poster' in Shaughnessy 2007: 248–71.

Ryan, J. (2002) 'Shakespeare's Globe Research Bulletin: Interviews with Company Members from the 2002 Theatre Season', 19, <http://www. shakespeares-globe.org/navigation/pdfs/2002%20Interviews.pdf>. Accessed 2 September 2004.

Rylance, M. (1997) 'Playing the Globe: Artistic Policy and Practice' in Mulryne & Shewring 1997: 169–76.

Schechner, R. (1982) 'Collective Reflexivity: Restoration of Behaviour', in Jay Ruby (ed.) *A Crack in the Mirror: Reflexive Perspectives in Anthropology*, Philadelphia: University of Pennsylvania Press, 39–81.

Schechter, J. (2003) 'Back to the Popular Source', in J. Schechter (ed.) *Popular Theatre: A Sourcebook*, London: Routledge, 3–11.

Schoch, R. (2002) *Not Shakespeare: Bardolatry and Burlesque in the Nineteenth Century*, Cambridge: Cambridge University Press.

Selbourne, D. (1982) *The Making of A Midsummer Night's Dream*, London: Methuen.

Shank, T. (2003) 'Political Theatre as Popular Entertainment: The San Francisco Mime Troupe', in J. Schechter (ed.) *Popular Theatre: A Sourcebook*, London: Routledge, 258–65.

Sharbaugh, P. (2006) 'Comedy: Elizabethantown', *Charleston City Paper*, 31 May.

Shaughnessy, R. (2000) (ed.) *Shakespeare in Performance: Contemporary Critical Essays*, Basingstoke: Palgrave Macmillan.

Shaughnessy, R. (2002) *The Shakespeare Effect: A History of Twentieth-Century Performance*, Basingstoke: Palgrave Macmillan.

Shaughnessy, R. (2007) (ed.) *The Cambridge Companion to Shakespeare and Popular Culture*, Cambridge: Cambridge University Press.

Shepherd, S. (2000) 'Acting against Bardom: Some Utopian Thoughts on Workshops', in Shaughnessy 2000: 218–34.

Sinfield, A. (2000) 'Royal Shakespeare: Theatre and the Making of Ideology', in Shaughnessy 2000: 171–93.

Solti, I. (2004) 'Playing with Fire: *The Jew of Malta* in a visible arena', research seminar at the University of Kent, 13 February 2004.

Southern, R. (1953) *The Open Stage*, London: Faber.

Staveacre, T. (1987) *Slapstick*, London: HarperCollins.

Steiner, G. (1961) *The Death of Tragedy*, London: Faber and Faber.

Talbot, P. (2003) interviewed by Steve Purcell in Eastbourne, 14 September.

Taylor, G. (1990) *Reinventing Shakespeare: A Cultural History from the Restoration to the Present*, London: Hogarth.

Taylor, G. (1999) 'Afterword: The incredible shrinking Bard', in Desmet & Sawyer 1999: 197–205.

Teeman, T. (2005) 'Rain or Shine, These Shows Go On', *Times*, 9 July.

Thomson, P. (2000) *On Actors and Acting*, Exeter: Exeter University Press.

Thomson, P. (2002) 'The Comic Actor and Shakespeare' in S. Wells & S. Stanton (eds) *The Cambridge Companion to Shakespeare on Stage*, Cambridge University Press, 137–54.

Tillyard, E. M. W. (1990) *The Elizabethan World Picture*, London: Penguin.

Tucker, P. (2002) *Secrets of Acting Shakespeare: The Original Approach*, London: Routledge.

Turner, C. (2004) 'Palimpsest or Potential Space? Finding a Vocabulary for Site-Specific Performance' in *New Theatre Quarterly* 20:4 (November 2004), 373–90.

Turner, V. W. (1969) *The Ritual Process: Structure and Anti-Structure*, New York: Aldine.

Turner, V. W. (1982) *From Ritual to Theatre: The Human Seriousness of Play*, New York: Performing Arts Journal.

Ubersfeld, A. (1981) *L'Ecole du Spectateur: Lire le Théâtre 2*, Paris: Editions Sociales.

Venturi, R. (1978) *Learning from Las Vegas: The Forgotten Symbolism of Architectural Form*, Cambridge, MA: MIT Press.

Wallace, N. (1995) 'Peter Brook, Theatre Space and the Tramway', Mulryne & Shewring 1995: 61–3.

Wayne, V. (2007) 'Shakespeare Performed: Kneehigh's Dream of *Cymbeline*', *Shakespeare Quarterly* 58:2, 228–37.

Webster, M. (2000) *Shakespeare Without Tears: A Modern Guide for Directors, Actors, and Playgoers*, Mineola, NY: Dover.

Weimann, R. (1987) *Shakespeare and the Popular Tradition in the Theater: Studies in the Social Dimension of Dramatic Form and Function*, Baltimore/London: Johns Hopkins University Press.

Weimann, R. (2000) *Author's Pen and Actor's Voice: Playing and Writing in Shakespeare's Theatre*, Cambridge: Cambridge University Press.

Weinberg, M. S. (2003) 'Community-Based Theatre: A Participatory Model for Social Transformation' in Kattwinkel 2003: 185–98.

Wells, S. (1965) 'Shakespearean Burlesques', *Shakespeare Quarterly* 16:1, 49–61.

Wells, S. (1977) (ed.) *Nineteenth-Century Shakespeare Burlesques*, 5 vols, London: Diploma Press.

Wells, S. (2005) 'Bold Adventures: the influence of William Poel', programme note for *The William Poel Festival*, London: Theatre Royal Haymarket.

Wesker, A. (1999) 'Theatre Cheats: An Open Letter to Trevor Nunn, in Two Acts', <http://www.arnoldwesker.com/>. Accessed 9 September 2007.

Wiles, D. (1998) 'The Carnivalesque in *A Midsummer Night's Dream*', in R. Knowles (ed.) *Shakespeare and Carnival: After Bakhtin*, Basingstoke: Palgrave Macmillian, 61–82.

Wiles, D. (2005) *Shakespeare's Clown: Actor and Text in the Elizabethan Playhouse*, Cambridge: Cambridge University Press.

Williams, D. (1985) '"A Place Marked by Life": Brook at the Bouffes du Nord', *New Theatre Quarterly*, 1, 39–74.

Williams, R. (1968) *Drama and Performance*, London: Watts.

Willis, S. (1991) *A Primer for Daily Life*, London: Routledge.

Wilmut, R. (1985) *Kindly Leave the Stage*, London: Methuen.

Wilson, J. D. (1918) 'The "Hamlet" Transcript, 1953', *The Library* 9, 217–47.

Wilson, J. D. (1936) (ed.) *Hamlet*, Cambridge: Cambridge University Press.

Worthen, W. B. (1997) *Shakespeare and the Authority of Modern Performance*, Cambridge: Cambridge University Press.

Worthen, W. B. (2003) *Shakespeare and the Force of Modern Performance*, Cambridge: Cambridge University Press.

Wright, L. B. (1926) 'Will Kemp and the Commedia Dell'Arte' *Modern Language Notes* 41: 8, 516–20.

Theatre productions

Aladdin (2004/5), dir. Sean Mathias, London: Old Vic Theatre.

As You Like It (1998), dir. Lucy Bailey, London: Shakespeare's Globe Theatre.

As You Like It (2003–5), dir. Peter Hall, UK and US tour: Peter Hall Company.

As You Like It (2004), dir. Simon Clark, UK tour: Chapterhouse Theatre.

As You Like It (2005), dir. David Lan, London: Wyndham's Theatre.

Bill Shakespeare's Italian Job (2003), dir. Malachi Bogdanov, Edinburgh: Cooperesque Productions Ltd & Gilded Balloon Productions.

The Bomb-itty of Errors (2000–), dir. Andy Goldberg, US and international tour: written and performed by Jordan Allen-Dutton, Jason Catalano, G. Q. and Erik Weiner.

Bouncy Castle Hamlet (2006), dir. William Seaward, Edinburgh: Strolling Theatricals.

The Comedy of Errors (1982), dir. Robert Woodruff, Chicago: Goodman Theater.

The Comedy of Errors (1987), dir. Robert Woodruff, New York: Lincoln Center Theater.

The Comedy of Errors (1990), dir. Ian Judge, Stratford-upon-Avon and London: Royal Shakespeare Company.

The Comedy of Errors (1999), dir. Pete Talbot, UK tour: The Rude Mechanical Theatre Company.

The Comedy of Errors (2005), dir. Nancy Meckler, Stratford-upon-Avon and London: Royal Shakespeare Company.

The Complete Works of William Shakespeare (abridged) (1987–), created by Jess Borgeson, Adam Long, and Daniel Singer, international: The Reduced Shakespeare Company. London: Criterion Theatre between 1996 and 2005.

Coriolanus (1991), dir. Michael Bogdanov, UK and international tour: English Shakespeare Company.

Cymbeline (2001), dir. Mike Alfreds, London: Shakespeare's Globe Theatre.

Cymbeline (2007), dir. Emma Rice, UK tour: Kneehigh Theatre.

Cymbeline (2007), dir. Declan Donnellan, international tour: Cheek by Jowl.

Five Day Lear (1999), dir. Tim Etchells, Sheffield: Forced Entertainment.

Hamlet (1965), dir. Peter Hall, Stratford-upon-Avon and London: Royal Shakespeare Company.

Hamlet (1980), dir. John Barton, Stratford-upon-Avon and London: Royal Shakespeare Company.

Hamlet (2000), dir. Giles Block, London: Shakespeare's Globe Theatre.

Hamlet (2000), dir. Peter Brook, international tour: Young Vic/Bouffes du Nord.

Hamlet (2007), dir. Elizabeth LeCompte, New York/international tour: The Wooster Group.

The Hamlet Project (2007–), dir. Tim Carroll and Tamara Harvey, London: The Factory.

Henry IV Part 1 (2005), dir. Nicholas Hytner, London: National Theatre.

Henry IV Part 2 (2005), dir. Nicholas Hytner, London: National Theatre.

Henry V (1997), dir. Richard Olivier, London: Shakespeare's Globe Theatre.

Henry VI Part I – The War Against France (2006), dir. Michael Boyd, Stratford-upon-Avon and London: Royal Shakespeare Company.

Henry VI Part II – England's Fall (2006), dir. Michael Boyd, Stratford-upon-Avon and London: Royal Shakespeare Company.

Henry VI Part III – The Chaos (2006), dir. Michael Boyd, Stratford-upon-Avon and London: Royal Shakespeare Company.

In Pursuit of Cardenio (2006), dir. Ken Campbell, Edinburgh: The Sticking Place:

Julius Caesar (2005), dir. David Farr, UK tour: Royal Shakespeare Company.

La Didone (2007), dir. Elizabeth LeCompte, New York/international tour: The Wooster Group.

The League of Gentlemen (2001), created by Jeremy Dyson, Mark Gatiss, Steve Pemberton, and Reece Shearsmith, London: Theatre Royal, Drury Lane.

Macbeth (1998–), dir. Ginger Perkins, Edinburgh: Frantic Redhead Productions.

Macbeth (1999/2000), dir. Gregory Doran, UK and international tour: Royal Shakespeare Company.

Macbeth (2004), dir. Max Stafford-Clark, UK and international tour: Out of Joint.

Macbeth (2006), dir. Paata Tsikurishvili, Washington DC/Arlington, Virginia: Synetic Theater.

Macbeth Re-Arisen (2006), dir. David Mence, Edinburgh: White Whale Theatre.

Macbeth the Panto (2004), dir. Andy Barrow, UK tour: Oddsocks Productions.

MacHomer (1996–), created by Rick Miller, international: Wyrd Productions.

Measure for Measure (2004/2006), dir. Simon McBurney, London/international tour: Complicite/National Theatre.

The Merchant of Venice (1970), dir. Jonathan Miller, London: Old Vic Theatre.

The Merchant of Venice (1991), dir. Tim Luscombe, UK and international tour: English Shakespeare Company.

The Merchant of Venice (1993), dir. David Thacker, Stratford-upon-Avon and London: Royal Shakespeare Company.

The Merchant of Venice (1998), dir. Richard Olivier, London: Shakespeare's Globe Theatre.

A Midsummer Night's Dream (1970), dir. Peter Brook, international tour: Royal Shakespeare Company.

A Midsummer Night's Dream (1991), created by Footsbarn, international tour: Footsbarn Theatre.

A Midsummer Night's Dream (2002), dir. Mike Alfreds, London: Shakespeare's Globe Theatre.

A Midsummer Night's Dream (2003), dir. Edward Hall, UK and international tour: Propeller.

A Midsummer Night's Dream (2006), dir. Tim Supple, international tour: Dash Arts in association with The British Council.

Much Ado About Nothing (2004), dir. Charlotte Conquest, Canterbury: Creation Theatre.

Much Ado About Nothing (2006), dir. Marianne Elliott, Stratford-upon-Avon and London: Royal Shakespeare Company.

Othello (2004), dir. Gregory Doran, Stratford-upon-Avon and London: Royal Shakespeare Company.

Pericles (2005), dir. Kathryn Hunter, London: Shakespeare's Globe Theatre.

The Pocket Dream (1992), dir. Pip Broughton, Nottingham/London: Nottingham Playhouse/Albery Theatre.

Richard II (1995), dir. Deborah Warner, London: National Theatre.

Richard II (2005), dir. Trevor Nunn, London: Old Vic Theatre.

Richard III (1990), dir. Richard Eyre, international tour: National Theatre.

Richard III (1992), dir. Barrie Rutter, Yorkshire: Northern Broadsides.

Richard III (2003), dir. Barry Kyle, London: Shakespeare's Globe Theatre.

Romeo and Juliet (2003), dir. Gísli Örn Gardarsson, London: Vesturport/Young Vic.

Romeo and Juliet (2004), dir. Tim Carroll, London: Shakespeare's Globe Theatre.

Romeo and Juliet (2006), dir. Steve Purcell, UK tour: The Pantaloons.

Shakespeare for Breakfast (2006), dir. Damien Sandys, Edinburgh: C Theatre.

The Shakespeare Revue (1994–5), dir. Christopher Luscombe and Malcolm McKee, London: Royal Shakespeare Company.

The Shakespeare Sketch (1989), dir. Stephen Fry, London: Sadler's Wells Theatre.

Shall We Shog? (2005), dir. Ken Campbell, London: Shakespeare's Globe Theatre.

The Storm (2005), dir. Tim Carroll, London: Shakespeare's Globe Theatre.
The Taming of the Shrew (2007), dir. Edward Hall, UK and international tour: Propeller.
The Tempest (1988), dir. Jonathan Miller, London: Old Vic Theatre.
The Tempest (2000), dir. Jonathan Kent, London: Almeida Theatre.
The Tempest (2004), dir. Oliver Gray, UK tour: Illyria.
The Tempest (2005), dir. Tim Carroll, London: Shakespeare's Globe Theatre.
Titus Andronicus (2006), dir. Lucy Bailey, London: Shakespeare's Globe Theatre.
Titus Andronicus (2006), dir. Yukio Ninagawa, international tour: The Ninagawa Company.
Twelfth Night (1971), dir. Anthony Tuckey, Liverpool: Liverpool Playhouse.
Twelfth Night (2001), dir. Pete Talbot, UK tour: The Rude Mechanical Theatre Company.
The Two Gentlemen of Verona (1996), dir. Jack Shepherd, London: Shakespeare's Globe Theatre.
The Two Gentlemen of Verona (2000), dir. Pete Talbot, UK tour: The Rude Mechanical Theatre Company.
The Two Gentlemen of Verona (2005), dir. Fiona Buffini, UK tour: Royal Shakespeare Company.
The Wars of the Roses (1963–4), dir. Peter Hall, Stratford-upon-Avon and London: Royal Shakespeare Company.
The Winter's Tale (1992), dir. Annabel Arden and Annie Castledine, international tour: Complicite.
The Winter's Tale (2005), dir. Steve Purcell, UK tour: The Pantaloons.
The Winter's Tale (2006), dir. Paul Burbridge, UK tour: Riding Lights.

Films

10 Things I Hate About You (1999), dir. Gil Junger, United States: Touchstone Pictures.
Blackadder: Back and Forth (1999), dir. Paul Weiland, United Kingdom: Millennium Dome/BSkyB/BBC.
Clueless (1995), dir. Amy Heckerling, United States: Paramount Pictures.
Dead Poets Society (1989), dir. Peter Weir, United States: Touchstone Pictures.
Get Over It (2001), dir. Tommy O'Haver, United States: Miramax Films.
Hamlet (1965), dir. John Gielgud, United States: Theatrofilm/Warner Bros.
Hamlet (1996), dir. Kenneth Branagh, UK/US: Columbia Pictures/Castle Rock Entertainment.
Henry V (1944), dir. Laurence Olivier, United Kingdom: Rank Film Distributors.
Hot Fuzz (2007), dir. Edgar Wright, United Kingdom: Universal Pictures.
In the Bleak Midwinter (1995), dir. Kenneth Branagh, United Kingdom: Castle Rock Entertainment.
A Knight's Tale (2001), dir. Brian Helgeland, United States: Columbia Pictures/ Twentieth Century Fox.
Last Action Hero (1993), dir. John McTiernan, United States: Columbia Pictures.
Moulin Rouge! (2001), dir. Baz Luhrmann, Australia/US: Twentieth Century Fox.
O (2001), dir. Tim Blake Nelson, United States: Lions Gate Films.
The Postman (1997), dir. Kevin Costner, United States: Warner Bros.

Robin Hood: Prince of Thieves (1991), dir. Kevin Reynolds, United States: Warner Bros.

Romeo + Juliet (1996), dir. Baz Luhrmann, United States: Twentieth Century Fox.

Shakespeare In Love (1998), dir. John Madden, UK/US: Universal Pictures/Miramax Films.

She's the Man (2006), dir. Andy Fickman, United States: DreamWorks.

Stage Beauty (2004), dir. Richard Eyre, UK/US: Lions Gate Films.

Titus (1999), dir. Julie Taymor, Italy/US: Twentieth Century Fox.

True Identity (1991), dir. Charles Lane, United States: Buena Vista Pictures.

Television broadcasts

Around the Beatles (1964), ITV, 6 May.

Big Train (1998/2002), BBC Two, 12 episodes.

The Black Adder, 'The Foretelling' (1983), BBC1, 15 June.

Dead Ringers (2002–7), BBC Two, 42 episodes.

Doctor Who, 'The Shakespeare Code' (2007), BBC One, 7 April.

Live From Lincoln Center: The Comedy of Errors (1987), PBS, 24 June.

Maid Marian and her Merry Men, 'They Came From Outer Space' (1993), BBC1, 28 January.

The Mark Thomas Comedy Product (1996–2002), Channel 4, 45 episodes.

Monty Python's Flying Circus (1969–74), BBC1, 45 episodes.

ShakespeaRe-Told, 'A Midsummer Night's Dream' (2005), BBC One, 28 November.

The Muppet Show, 'Peter Sellers' (1978), ITV, 1 January (UK)/CBS, 24 February (US).

The Muppet Show, 'Christopher Reeve' (1980), ITV/CBS, 7 February.

The Music of Lennon and McCartney (1965), ITV, 16 December.

The Simpsons, 'Tales from the Public Domain' (2002), Fox, 17 March.

Whose Line Is It Anyway?, Season 3, Episode 14 (1991), Channel 4, 19 April.

Radio broadcasts

Hamlet Part II (1992), BBC Radio 3, 27 April.

Lenny and Will (2006), BBC Radio 4, 25 March.

Audio recordings

All citations from Shakespeare are from the Taylor & Wells edition (1986), Oxford: Clarendon Press.

Luscombe, C. & McKee, M. (1995) *The Shakespeare Revue* (CD), London: Jay Productions Ltd.

Newspaper reviews of theatre, television, and film are not listed in the bibliography: where cited, newspaper and date of publication are listed in the main body of the text. Interviews and other newspaper articles are listed in the bibliography.

Sellers, P. (1965) *A Hard Day's Night* (7-inch vinyl), London: Parlophone Records.

Spellings have been standardised to British.

Unless indicated otherwise, quotations from television broadcasts, films, or CDs were transcribed directly from recordings.

Williams, R. (1979) *Reality ... What A Concept* (CD), Los Angeles: Laugh.com Comedy Recording Series (reissue).

Williams, R. (1983) *Throbbing Python of Love* (CD), Los Angeles: Laugh.com Comedy Recording Series (reissue).

Index

CPSIA information can be obtained at www.ICGtesting.com
Printed in the USA
BVOW05*1226250416

445497BV00017B/302/P